Synthese Library

Studies in Epistemology, Logic, Methodology, and Philosophy of Science

Volume 400

The aim of *Synthese Library* is to provide a forum for the best current work in the methodology and philosophy of science and in epistemology. A wide variety of different approaches have traditionally been represented in the Library, and every effort is made to maintain this variety, not for its own sake, but because we believe that there are many fruitful and illuminating approaches to the philosophy of science and related disciplines.

Special attention is paid to methodological studies which illustrate the interplay of empirical and philosophical viewpoints and to contributions to the formal (logical, set-theoretical, mathematical, information-theoretical, decision-theoretical, etc.) methodology of empirical sciences. Likewise, the applications of logical methods to epistemology as well as philosophically and methodologically relevant studies in logic are strongly encouraged. The emphasis on logic will be tempered by interest in the psychological, historical, and sociological aspects of science.

Besides monographs *Synthese Library* publishes thematically unified anthologies and edited volumes with a well-defined topical focus inside the aim and scope of the book series. The contributions in the volumes are expected to be focused and structurally organized in accordance with the central theme(s), and should be tied together by an extensive editorial introduction or set of introductions if the volume is divided into parts. An extensive bibliography and index are mandatory.

More information about this series at http://www.springer.com/series/6607

Ilkka Niiniluoto

Truth-Seeking by Abduction

Springer

Ilkka Niiniluoto
University of Helsinki
Helsinki, Finland

Synthese Library
ISBN 978-3-030-07571-2 ISBN 978-3-319-99157-3 (eBook)
https://doi.org/10.1007/978-3-319-99157-3

This Springer imprint is published by the registered company Springer Nature Switzerland AG
The registered company address is: Gewerbestrasse 11, 6330 Cham, Switzerland

Preface

The *Oxford English Dictionary* defines the verb "to abduct" as "to take (a person) away by force or deception, or without the consent of his or her legal guardian; to kidnap." Derived from the Latin words *ab* (away) and *duco* (to lead), abduction in popular culture refers also to aliens who take humans to their flying saucers. The origins of this usage, in English since the 1830s, goes back to the Athenian law where *apagoge* in Greek ("a leading away") meant arresting or taking a criminal before the proper magistrate.

As a philosophical term in English, "abduction" was used already in the seventeenth century as a translation of the Latin term abductio in Giulio Pacio's 1597 translation of Aristotle's *Analytica Priora.* Charles S. Peirce adopted this term in 1898 from Renaissance logicians and gave it a significant place in his philosophical system. Peirce (1839–1914) was the founder of American pragmatism, well-known and appreciated for his studies in formal logic and probabilistic reasoning. He argued already in 1865 that, besides deduction and induction, there is a third type of inference which he called "hypothesis," "presumption," "retroduction," or "abduction." This is reasoning from effects to causes or from surprising observations to explanatory theories. After Peirce's death, his ideas gradually become known with the publication of six volumes of his *Collected Papers* (*CP*) in 1931–1935.

Peirce's conception of abductive reasoning became a hot topic in the philosophy of science after World War II, when N. R. Hanson suggested that abduction is a logic of discovery, Gilbert Harman argued that all types of inductive reasoning can be reduced to inference to the best explanation (IBE), and Howard Smokler analyzed abduction as an important method of confirmation. Abductive inference became soon a fashionable theme in logic, artificial intelligence (AI), and semiotics. Illustrations and examples of abduction have been sought in everyday life, detective stories, and many scientific disciplines from astronomy to medicine.

I have been an admirer of Peirce ever since I read Justus Buchler's selection of his philosophical writings in 1972 when I was writing my PhD thesis on inductive logic. I discussed his ideas on abduction in an essay "Notes of Popper as Follower of Whewell and Peirce" written in 1975 (see Niiniluoto, 1978), and in my later work on fallibilism, scientific progress, and truthlikeness, I have treated Peirce as an inspiring background figure of critical scientific realism.

A book on abduction inevitably has to engage in the scholarly question of what Peirce really meant by this notion (see Chap. 1). But ideas have a life of their own, and the best tribute to the great American philosopher is to see what kinds of historical and systematic applications his notion of abduction and its variations may have. As Peirce himself noted in 1903, there are "pedants and pedagogues who are continually endeavoring to set up a sort of magistrate over thoughts and other symbols." Still, "the effort of all should be to keep the *essence* of every scientific term unchanged and exact." Following in this spirit "the ethics of terminology," in our survey, I also criticize some recent developments of the theory of abduction.

In Chap. 2, I discuss a topic which was neglected by Peirce: the method of analysis and synthesis, with its implication to one of Peirce's own favorite themes—detective stories. Chapter 3 outlines some logical and AI approaches to abduction. Chapter 4 gives examples of retroductions and abductions by studying various kinds of inverse problems in medicine, biology, and human sciences, and then follows abduction as discovery and pursuit (Chap. 5). The probabilistic Bayesian framework is used in the analysis of abduction and confirmation (Chap. 6) and inference to the best explanation (Chap. 7). In Chap. 8, abduction is reformulated as inference which leads to the truthlikeness of its conclusion. The final Chap. 9 summarizes the crucial significance of abduction for debates on scientific realism.

This book is not a collection of essays, but written as an integrated monograph. But I have used in the text passages from some of my earlier articles: "Abduction and Geometrical Analysis," in L. Magnani, N. Nersessian, and P. Thagard (eds.), *Model-Based Reasoning in Scientific Discovery* (New York: Kluwer and Plenum, 1999); "Truth-Seeking by Abduction," in F. Stadler (ed.), *Induction and Deduction in the Sciences* (Dordrecht: Kluwer, 1999); "Abduction and Truthlikeness," in R. Festa, A. Aliseda, and J. Peijnenburg (eds.), *Confirmation, Empirical Progress, and Truth-Approximation* (Amsterdam: Rodopi, 2005); "Structural Rules for Abduction," *Theoria* 22 (2007); "Abduction, Tomography, and Other Inverse Problems," *Studies in History and Philosophy of Science* 42 (2011); "Approximation of Laws," in G. Guo and C. Liu (eds.), *Scientific Explanation and Methodology of Science* (Singapore: World Scientific, 2014); "Unification and Confirmation," *Theoria* 31 (2016).

As the reader can see in the references, the preparation of this book has taken some decades. For the actual writing in 2017, I am grateful to the facilities and research environment provided for a professor emeritus by the Department of Philosophy, History, Culture, and Art Studies at the University of Helsinki. My intellectual debt to my most important teacher, the late Professor Jaakko Hintikka, is evident in this work. I have benefited from discussions about abduction with several colleagues, especially Atocha Aliseda, Gustavo Cevolani, Theo Kuipers, Sami Paavola, and Gerhard Schurz. As always, special thanks to my family for their patience and encouragement.

Helsinki, Finland Ilkka Niiniluoto

Contents

Chapter 1
Peirce on Abduction

Abstract As a philosophical term in English, 'abduction' was originally a seventeenth century translation of a Latin term used by Renaissance logicians. It was adopted in 1898 by Charles S. Peirce who gave it a significant role in his system. Peirce (1839–1914) was the founder of American pragmatism, well-known and appreciated for his studies in formal logic and probabilistic reasoning. He argued already in 1865 that, besides deduction and induction, there is a third type of inference which he called 'hypothesis', 'presumption', 'retroduction', or 'abduction'. This is reasoning from effects to causes or from surprising observations to explanatory theories. Peirce's account of this ampliative inference changed in important ways during the 50 years between 1865 and 1914. After Peirce's death, his ideas gradually become known with the publication of six volumes of his *Collected Papers* (*CP*) in 1931–1935. Peirce's conception of abductive reasoning became a hot topic in the philosophy of science after World War II, when N. R. Hanson suggested that abduction is a logic of discovery, Gilbert Harman argued that all types of inductive reasoning can be reduced to inference to the best explanation (IBE), Howard Smokler analyzed abduction as an important method of confirmation, and Larry Laudan treated abduction as an instruction for the pursuit of a hypothetical theory. In this chapter, we shall follow Peirce's steps in discussing abduction by analyzing its logical and probabilistic forms (Sects. 1.1, 1.2, and 1.3), its main interpretations (Sect. 1.4), and some current debates about Peirce's distinctions (Sect. 1.5).

Is there a third type of inference besides deduction and induction? Does this third type of inference play a significant role within scientific inquiry? A positive answer to both of these questions was advocated by Charles S. Peirce throughout his career, even though his opinions changed in important ways during the 50 years between 1865 and 1914. Peirce called the third kind of inference "hypothesis", "presumption", "abduction", or "retroduction".[1] In this chapter, we shall follow

[1] References to Peirce's works are given in the standard way: (*CP* x.y) refers to volume x and paragraph y of *Collected Papers* (Peirce 1931–1935 and 1958), (*W*) to *Writings*, (*RTL*) to *Reasoning and the Logic of Things* (Peirce 1992a), (*EP* 1) and (*EP* 2) to *The Essential Peirce* (1992b, 1998).

I. Niiniluoto, *Truth-Seeking by Abduction*, Synthese Library 400,
https://doi.org/10.1007/978-3-319-99157-3_1

Peirce's steps in discussing abduction by analyzing its logical form (Sects. 1.1, 1.2 and 1.3), its main interpretations (Sect. 1.4), and some debates about Peirce's distinctions (Sect. 1.5).

1.1 Classification of Inferences

Peirce made a distinction between three modes of scientific reasoning already in his Harvard Lectures during the spring of 1865. Starting from Aristotle's doctrine of induction as the inference of the major premise of a syllogism (of the first figure), Peirce observed that there is "a large class of reasonings" that are neither deductive nor inductive: reasoning *a posteriori* to a physical hypothesis, or inference of a cause from its effect (*W* 1:180). This reasoning, which Peirce called Hypothesis, can be represented as the inference of the minor premise of a syllogism. Besides Aristotle, Peirce's inspiration and the term 'hypothesis' came from the 1800 edition of Immanuel Kant's *Logic* (*W* 1:451).[2] This classification of inferences was elaborated in Peirce's Lowell Lectures in the fall of 1866, and published in the next year. In Peirce's series for *Popular Scientific Monthly* 10 years later, this distinction was presented in the article "Deduction, Induction, and Hypothesis" (1878) and later treated in detail in "A Theory of Probable Inference" (1883).

Peirce's starting point was the observation that a paradigm example of *deduction*, the Barbara syllogism of the first figure, can be inverted in two different ways. In Aristotle's own formulation, the two premises and the conclusion of Barbara are affirmative universal sentences of the form "G belongs to every F" (e.g. "All men are mortal"), but the inference is valid also when the minor premise is singular (e.g. "Socrates is a man"). Using modern logical notation with ∴ as a general sign for inference and allowing singular terms, the latter case of Barbara looks as follows:

(1) $(\forall x)(Fx \rightarrow Gx)$
Fb
$\therefore Gb.$

Induction is the inference of the major premise (rule) from the minor premise (case) and the conclusion (result):

(2) Fb
Gb
$\therefore (\forall x)(Fx \rightarrow Gx).$

[2]In 1868 Peirce cited, besides Kant, several authors who have supported the view of hypothesis as "an argument from consequence and consequent to antecedent": Chauvin, Newton, Hamilton, Mill, Herbart, and Beneke (*W* 2:219; *EP* 1:34–35).

Hypothesis is the inference of the minor premise from the major premise and the conclusion:

(3) $(\forall x)(Fx \rightarrow Gx)$
Gb
$\therefore Fb.$

Thus, hypothesis leads from the rule and the result to the case (see *CP* 2.623.) Simple variants of these schemata allow the case to refer to more than one instance b. Then the inversion (2), which Peirce called "crude induction" (*CP* 2.757, 6.473), expresses a typical inductive generalization based on several instances:

(4) $Fb_1 \& \ldots \& Fb_n$
$Gb_1 \& \ldots \& Gb_n$
$\therefore (\forall x)(Fx \rightarrow Gx).$

The difference between induction and hypothesis was described by Peirce in the following way:

> By induction, we conclude that facts, similar to the observed facts, are true in cases not examined. By hypothesis, we conclude the existence of a fact quite different from anything observed, from which, according to known laws, something observed would necessarily result. The former, is reasoning from particulars to the general law; the latter, from effect to cause. (*CP* 2.536.)

Thus, hypothetical reasoning (3) proceeds from effects to causes, while a typical deduction (1) proceeds from causes to effects. The latter requirement is Aristotle's well-known condition for scientific syllogisms. For example, inference from fire to smoke can be formulated as a deduction (1), while the inverse inference from smoke to fire is a hypothesis (3).

Peirce also stated, already in 1865, that hypothesis is an inference to an *explanation*, and the resulting inference (1) is then an "explanatory syllogism" (*W* 1:267, 428, 440). The same holds of examples where Barbara involves only universal statements:

(5) $(\forall x)(Fx \rightarrow Gx)$
$(\forall x)(Bx \rightarrow Fx)$
$\therefore (\forall x)(Bx \rightarrow Gx).$

In this case, the hypothetical inversion of (5) takes the form

(6) $(\forall x)(Fx \rightarrow Gx)$
$(\forall x)(Bx \rightarrow Gx)$
$\therefore (\forall x)(Bx \rightarrow Fx).$

For example, the supposition that light (B) is ether waves (F) explains why light gives certain peculiar fringes or is polarizable (G), given the law that ether waves give these fringes (*W* 1:267). In this way hypothetic inference gives "scientific understanding" of a singular or general fact by providing the minor premise for its deductive explanation (cf. Aristotle 1989, 223).

Hence, Peirce's account of deduction covers what Carl G. Hempel in 1948 called the deductive-nomological (DN) explanation of singular facts (1) and the DN explanation of laws (5) (see Hempel 1965). Peirce's statement that "a fact is *explained* when a proposition – possibly true – is brought forward, from which that fact follows syllogistically" (*W* 1:425, 440, 452) agrees with John Stuart Mill's view in *A System of Logic* (first edition in 1843) that explanation of facts and uniformities in nature means subsumption under laws or deduction from laws (see Mill 1906, 305). Karl Popper (1957), 125, argued that Mill had not realized the role of specific initial conditions besides universal laws, but Peirce's syllogistic formulations of explanations (1) and (5) avoids this objection. Peirce's account of hypothesis, as an inversion of deduction, covers *singular abduction* (3) and *general abduction* (6).[3]

In his 1878 paper, Peirce gave three examples of hypothetical inference (*CP* 2.625). In meeting a Turkish man surrounded by four horsemen holding a canopy over his head, Peirce inferred that this man was the governor of the province. In finding fish-like fossils far in the interior of the country, we suppose that the sea once washed over this land. Countless documents and monuments refer to a conqueror called Napoleon Bonaparte, and we explain them by supposing that Napoleon really existed.

These examples illustrate the fact that inference from effect to cause may involve reasoning backwards in time. This is the motive why in 1896 Peirce called hypothesis *retroduction* (see *RLT*).[4] As Peirce himself knew, such retroductions are typical in solving crimes both in real life and in detective stories (see Chap. 2 below).

What Peirce did not observe, however, was that hypothetical inference (proceeding from effects to causes) and explanatory deductions (proceeding from causes to effects) correspond to what medieval and Renaissance Aristotelians called *resolutio* and *compositio*, respectively. These Latin terms were translations of the Greek terms *analysis* and *synthesis*, as used in geometry, so that Peirce's abduction has an interesting historical connection to the geometrical method of analysis (see Chap. 2 below).

The example of Napoleon Bonaparte also illustrates the important point that hypothetical reasoning "infers very frequently a fact not capable of direct observation" (*CP* 2.641):

> The great difference between induction and hypothesis is, that the former infers the existence of phenomena such as we have observed in cases which are similar, while hypothesis supposes something of a different kind from what we have directly observed, and frequently something which it would be impossible for us to observe directly. (*CP* 2.640.)

[3]Cf. the distinction between factual abduction and law abduction in Schurz (2008a). A variant of (3), where the singular conclusion Fb is replaced by an existential statement (Ey)Ryb, is called "first-order existential abduction" by Schurz (2008a). For example, when a man dies with an injury, the police suspects that someone has killed him. For similar ideas in AI research, see Josephson and Josephson (1994); Flach and Kakas (2000).

[4]Hempel used the term 'retrodiction' in 1958 without reference to Peirce. See Hempel (1965, 173).

Thus, besides "horizontal" inferences leading to observable facts, hypothetical reasoning is typically "vertical".[5] In his 1898 Cambridge lectures, Peirce even claimed that inference of a *vera causa*, as a "state of things known to be present", is not really hypothetical but inductive (*EP* 2:54).

To keep the number of different kinds of inferences as three, in his early papers Peirce treated inference by *analogy* as a kind of hypothetical inference: "we find that in certain respects two objects have a strong resemblance, and infer that they resemble one another strongly in other respects" (*CP* 2.624). In 1883, he formulated a probabilistic argument which clearly involves analogical reasoning (*CP* 2.706). He classified this inference as being a hypothesis, but noted that it has also been treated as "an induction of characters".[6] Peirce illustrated this kind of reasoning with an example where two torn pieces of paper with anonymous writing fit together, and it is inferred that a certain person is the author of these notes (*CP* 2.632). This can be viewed as an inference to a common cause, which explains the coexistence of two apparently separate but similar things.[7]

1.2 Probabilistic Reasoning

Peirce's first formulation of induction and hypothesis as probable arguments in 1867 was inspired by John Venn's frequentist treatment of probability in *The Logic of Chance* (1866). In his 1878 paper, Peirce replaced the universal premise 'All F are G' in (1) by the statistical statement 'Most F are G'. Thus, (1) is replaced by *probable deduction*:

(7) Most of the beans in this bag are white.
These beans are from this bag.
∴ Probably, most of this handful of beans are white.

Its two inversions again lead to induction:

(8) Most of this handful of beans are white.
These beans are from this bag.
∴ Probably, most of the beans in this bag are white.

[5]For the horizontal – vertical distinction, see Psillos (1996). A similar distinction between observable-fact-abduction and unobservable-fact-abduction is made by Schurz (2008a). Besides past events, unobservables may include theoretical entities in science: an observed curve in Wilson's cloud chamber can be abductively explained as a trace of an electron.

[6]Cf. the discussion of J.S. Mill's notion of analogy in Niiniluoto (1988).

[7]See Salmon (2001a), 74. Schurz (2008a) calls this kind of inference "hypothetical common-cause abduction", which is a species of "second-order existential abduction". Such abductions may seek explanations for empirical regularities by postulating microstructures and hidden causal mechanisms. Minnameier (2107) claims that common-cause inferences are no abductions, but we disagree by giving an account of different kinds of theoretical unification in Chap. 6.

and to hypothesis:

(9) Most of the beans in this bag are white
Most of this handful of beans are white.
∴ Probably, these beans are from this bag.

If induction (8) is rewritten by replacing 'most' with 'few', the inferential pattern can be obtained from (7) by the same transformation (denying the result, accepting the case, and denying the rule) which leads from the Barbara syllogism (1) to the Bocardo syllogism of the third figure:

(10) ~Gb
Fb
∴ (Ex)(Fx & ~Gx).

Induction is thus probabilistic reasoning in the third figure (*CP* 2.627). The transformation (denying the result, accepting the rule, denying the case) leads from Barbara (1) to the Baroco syllogism in the second figure:

(11) ~Gb
(x)(Fx → Gx)
∴~Fb

A similar transformation leads from the probable deduction (7) to

(12) Few beans of this handful are white.
Most beans in this bag are white.
∴ Probably, these beans were taken from another bag.

Thus, hypothetical inference is probabilistic reasoning in the second figure (see also *RLT*). About the schemata (9) and (12) Peirce made the interesting additional point:

> When we adopt a certain hypothesis, it is not alone because it will explain the observed facts, but also because the contrary hypothesis would probably lead to results contrary to the observed. (*CP* 2.628.)

So here hypothetical reasoning is not only inference to an explanation but what Harman (1965) called "inference to the best explanation" (IBE).

In his 1883 article, Peirce further formalized probabilistic arguments by replacing the major premise by a precise statistical statement 'The proportion r of Fs are G'. Then we obtain the schema of *simple probable deduction* as a variant of deduction (1):

(13) The proportion r of the Fs are G;
b is an F;
It follows, with probability r, that b is a G.

(*CP* 2.695). Here the conclusion can be taken to be 'b is a G', and the probability r indicates "the modality with which this conclusion is drawn and held to be true" (*CP* 2.720). More precisely, the probability of a mode of argument is its *truth-frequency*, i.e., its ability to yield true conclusions from true premises. For example, this applies to the probability r in (13). For ordinary deduction, which is necessarily truth-preserving, the truth-frequency is one.

Peirce generalized (13) to the schema of *Statistical Deduction*:

(14) The proportion r of the Fs are G,
b′, b″, b‴, etc. are a numerous set, taken at random from among the Fs;
Hence, probably and approximately the proportion r of the b's are G.[8]

What Peirce called "quantitative induction" can now be obtained by inverting the argument (14): the premises state the observed proportion of Gs in a random sample of Fs and the conclusion states the proportion of Gs in the whole population of Fs.

(15) b′, b″, b‴, etc. are a numerous set, taken at random from among the Fs,
the proportion r of the b's are G;
Hence, probably and approximately the proportion r of the Fs are G.

This inference, often called statistical generalization, and later discussed by Hans Reichenbach in his "straight rule", includes inductive generalization (4) as its special case ($r = 1$). The hypothetical inversion of (14) would correspond to the inference (9). Its special case, inversion of (13) leads to simple probabilistic abduction:

(16) The proportion r of the Fs are G.
Gb
∴ Probably, Fb.

This is a generalization of the singular hypothetical inference (3).

Peirce's treatment of probabilistic reasoning anticipated important twentieth century frequentist accounts of induction like Reichenbach's probability logic and the Neyman-Pearson theory of statistical tests. If Peirce's truth-frequencies are changed to epistemic probabilities, his model of probable deduction (13) corresponds to what Rudolf Carnap (1962) and Isaac Levi (1967) have called "direct inference". This type of inference has often been discussed as a model of prediction. But, in his discussion of probabilistic inference, Peirce repeated the point that hypotheses are adopted "for the sake of the explanation", so that (13) is an explanatory statistical syllogism (*CP* 2.716). Thus, Peirce here anticipated C. G. Hempel's 1962 model of inductive-probabilistic explanation of particular facts. Moreover, his schema (14) for statistical deduction is richer than Hempel's IP model and Salmon's SR model, since it provides an account of probabilistic explanation of statistical facts (see Niiniluoto 1981). If the major premise of (13) is interpreted as a probabilistic law with a dispositional "would-be" or propensity, as Peirce suggested in 1910 (*CP* 2.666), his schema corresponds to the propensity model of IP explanation (see Fetzer 1981; Niiniluoto 1981).

In Niiniluoto (1993a), I argued that Wesley Salmon has ignored Peirce's historical role in the theory of probabilistic explanation. Hempel conceded the point (see

[8]Peirce calls schemas (13) and (14) "deductions", as they proceed from a population to a sample, but from the logical point of view they are probabilistic arguments.

Niiniluoto 2000), but it seems that Salmon remained unconvinced. In his papers Salmon (2001a, b), which were based on a conference in Haifa in 1998 but appeared after his death in April 2001, Salmon presented a Bayesian critique of inference to the best explanation. Peirce is not mentioned at all. The history of DN explanation is traced to Aristotle, Mill, and Popper, and the history of abduction starts with Harman (1965). When Salmon argues that "we cannot fairly evaluate inference to the best explanation without considering statistical models of explanation" (Salmon 2001a, 72), he is in fact repeating a point made forcefully by Peirce already in 1878.

1.3 Abduction Reconsidered

Peirce's early classification of inferences was based upon Aristotelian syllogisms. However, a complete characterization of logical reasoning cannot be found in this way, since it is able to give only typical simple examples of some kinds of inferences. Different logical forms of valid deductions cannot be given by any finite list. Peirce realized this quite well when he developed his theory of relations in the 1870s and his theory of quantification in the 1880s. He did not construct an axiomatic treatment of deduction, as logicians from Frege and Russell to Hilbert and Gentzen. Instead, he gave a general characterization of validity in semantical or model-theoretical terms. In "On the Algebra of Logic" (1880), Peirce introduced a statement of the form $P \preccurlyeq C$ to express "the truth of the leading principle" of the inference $P \therefore C$. This statement signifies that "every state of things in which a proposition of the class P is true is a state of things in which the corresponding propositions of the class C are true" (*EP* 1:203).[9] Thus, the characteristic of valid deduction is that it is necessarily truth-preserving.

For Peirce inference is a voluntary act which aims to generate new knowledge. Inference leads to a judgment by following a general method of reasoning (cf. Kapitan 1997, 479). Peirce divided inferences into *analytic* (deductive) and *synthetic* (induction and hypothesis) (*CP* 2.624). Analytic inferences are "explicative", while synthetic inferences are "ampliative". Peirce knew that it is only a "half-truth" to claim with Kant that deduction merely explicates what is implicitly asserted in the premises (*EP* 2:96). Arguing that deduction proceeds by diagrams, Peirce made a distinction between *corollarial* and *theorematic* deduction: the former finds the conclusion by observing the diagram, but the latter also "performs an ingenious experiment upon the diagram" (*EP* 2:298). This distinction corresponds to Jaakko Hintikka's division between trivial and non-trivial deductions (or surface tautologies

[9] An exact model-theoretical characterization of logical consequence was given by Alfred Tarski in 1935, but his definition was anticipated by Bernard Bolzano in 1837 (see Bolzano 1972, 209).

and depth tautologies), where the latter require the introduction or construction of new individuals or objects in the proof (see Hintikka 1997).[10]

According to Peirce's pragmatist theory of truth, "the opinion which is fated to be ultimately agreed to by all who investigate, is what we mean by truth, and the object represented in this opinion is the real" (*CP* 5.407). This does not mean that truth and reality are constituted by the process of inquiry, since "the opinion which would finally result from investigation does not depend on how anybody may actually think" (*CP* 5.408). Peirce is a realist in the sense that our hypotheses here and now have truth values, even though his fallibilism acknowledges we can never be absolutely certain about such truths in any special case (*CP* 5.311). Thus, a claim like "truth is that concordance of an abstract statement with the ideal limit towards which endless investigation would tend to bring scientific belief" (*CP* 5.565) should not be understood as a *definition* of truth but rather as an expression of the idea that the method of science is able to *approach to the truth* in the long run.[11] In his Cambridge lectures in 1898, Peirce stated that one of the most wonderful features of reasoning is its tendency "to correct itself, and the more so the more wisely its plan is laid" (*CP* 5.575; *EP* 2:43). In particular, as an ampliative inference induction is fallible, but by its self-corrective nature it is guaranteed to approach the truth in the long run. According to Peirce, induction can be justified by the fact that it "pursues a method which, if duly persisted in, must, in the very nature of things, lead to a result indefinitely approximating to the truth in the long run" (*CP* 2.781).

Peirce's deeper insights about deduction and scientific method led him to revise his views concerning induction and hypothesis as well.[12] In his manuscript "Lessons from the History of Science" (c. 1896) he introduced the term *retroduction* for "the provisional adoption of a hypothesis" (*CP* 1.68). Already in his unfinished

---

[10]Some illustrations of theorematic reasoning in geometry are given in Chap. 2. Ketner (1985) interprets Peirce's diagrammatic methods in an overly empiricist way, where diagrams are literally observed instead of imagined (cf. Peirce 1976, vol. 4, 38), with the implausible conclusion that for Peirce mathematics is an empirical, observational and hypothesis-confirming science. This account loses the sense in which reasoning by diagrams is a form of deduction (cf. Pietarinen 2006). For the relevance of these debates to some recent accounts of abduction, see Sect. 2.3.

[11]One of Peirce's lectures in 1903 is entitled "Pragmatism as the Logic of Abduction" (see *EP* 2:226–241). Peirce's characterization of truth as the limit of inquiry motivates the methodological role of abduction as truth-seeking. But we shall see in this book that abduction has a significant function also within scientific realism, which accepts the correspondence theory of truth (see Niiniluoto 1984, Ch. 5; 1999a).

[12]Abduction is an important concept also in Peirce's semiotics, which includes arguments as a special class of signs (see Short 2007). Peirce's distinction between icon, index, and symbol on the basis of the relation of sign to its object is also relevant (*CP* 4.536). In his unfinished "Minute Logic" Peirce associates abductive conclusion to icons which resemble their objects (*EP* 2:287, see Kruse 1986). It seems also natural to link abduction, as inference of a cause from its effect, to indices where the relation between a sign and its object is causal: for example, a detective reasons from the clues to a murderer and a medical doctor from the symptoms to a disease (cf. Sect. 2.4). Abductions as hypotheses seeking general true laws are symbols (*EP* 2:287). These three levels correspond to Peirce's categories of Firstness (quality), Secondness (impact), and Thirdness (generality).

"Dictionary of Logic" in 1867 Peirce had noted that modern logicians have translated Aristotle's account of *apagoge* in *Prior Analytics*, II, Chapter 25 (69a20-36) into Latin as *abductio* (*W* 2:108), so that *abduction* is the English counterpart of the term used by Julius Pacius in 1597. This notion is sometimes translated as "reduction", but Peirce denies that it has any connection to indirect or "apagogical proof". Aristotle's example refers to a syllogism where the major premise is evident but the minor only probable. So, for example, it is known that knowledge (B) is teachable (A), but not whether virtue (C) is teachable. But if 'C is B' is more credible than 'C is A', then we come nearer to knowing that 'C is A' by the syllogism
(1) B is A
(2) C is B
(3) ∴ C is A.

In this early stage Peirce did not associate "abduction" to his account of hypothetical inference. But later he came to think that Aristotle's text here is corrupted (*CP* 1.65), and what Aristotle really meant was the inversion given by his own schema of hypothesis (3) (cf. *CP* 5.144, 7.249, *EP* 2:527–528).[13] For this reason he was ready in 1898 to use this "otherwise useless term" *abduction* for his new account of hypothesis.

In his Cambridge Lectures of 1898, Peirce said that induction is "probable reasoning in the third figure"; this is what Aristotle called *epagoge* (*RLT*, 139), translated as *inductio* ("a leading in") by Cicero. Hypothesis or retroduction, on the other hand, is "probable reasoning in the second figure". This is what Aristotle called *apagoge* ("a leading out"). The general form of such *abduction* is

(17) If μ were true, π, π', π'' would follow as miscellaneous consequences.
But π, π', π'' are in fact true.
∴ Provisionally, we may suppose that μ is true.

(*Ibid*., 140.) It is also called "adopting a hypothesis for the sake of its explanation of known facts".

In his 1903 Harvard lectures, Peirce expressed the general form of the "operation of adopting an explanatory hypothesis" as follows (*CP* 5.189, *EP* 2:231):

(18) The surprising fact C is observed;
But if A were true, C would be a matter of course.
Hence, there is reason to suspect that A is true.

This schema, which has become Peirce's best known or canonical formulation of abduction, indicates how a hypothesis can be "abductively conjectured" if it accounts "for the facts or some of them". Peirce added that the conclusion is not A itself, but the assertion that "there is reason to suspect that A is true". In this sense, abduction "only infers a *may-be*" from an actual fact (*CP* 8.238).

[13]According to Anderson (1986), 146, Peirce's view differs from the usual interpretation of Aristotle's *apagoge*, where 'C is A' is uncertain, since "the conclusion of an abduction is known as a fact". This is correct, but here Anderson refers by "abduction" to the explanatory syllogism.

Schema (18) is obviously a generalization of the original patterns (3) and (6) of hypothetical inference: the emphasis that the fact C is surprising (and therefore in need of explanation) has been added, and there are no restrictions on the logical complexity of A. As here A may be a general theory, it might be said to express *theoretical abduction* (in contrast to singular abduction (3)).[14] The idea of explanation is maintained in the second premise, but this is not any more associated with the relation of cause and effect.

For Peirce, the motive in proposing explanations is to advance our knowledge from its present state to a higher stage: if we already know that C, and a hypothesis A explains why C, then A is a good candidate for being true as well. Peirce is a realist in the sense that for him explanatory hypotheses are candidates for truth. The second premise of (18) clearly expresses the idea that A is a potential explanation of the fact C,.[15] In this view, finding an explanation of known facts and regularities is a fallible or non-demonstrative method of *truth-seeking*. This attempt to find new truths is clearly the main function of Peirce's abduction. In other words, abduction has a crucial role in learning (Nesher 2001) and scientific progress (Niiniluoto 1984).

Already in 1878 Peirce argued that hypotheses have to be put into "fair and unbiased" tests by comparing their predictions with observations (*CP* 2.634). Later he claimed that he had earlier misunderstood the relation between induction and hypothesis (*CP* 8.227). In his papers and lectures in 1901–1903, Peirce defined induction in a new way as "the operation of testing a hypothesis by experiment" (*CP* 6.526), which is quite different from (4) or (15), whereas abduction is an "inferential step" which is "the first starting of a hypothesis and the entertaining of it, whether as a simple interrogation or with any degree of confidence" (*CP* 6.525). Here abduction and induction are successive steps in scientific inquiry:

> Abduction, on the other hand, is merely preparatory. It is the first step of scientific reasoning, as induction is the concluding step... The method of either is the very reverse of the other's. Abduction makes its start from the facts, without, at the outset, having any particular theory in view, though it is motivated by the feeling that a theory is needed to explain the surprising facts. Induction makes its start from a hypothesis which seems to recommend itself, without at the outset having any particular facts in view, though it feels the need of facts to support the theory. Abduction seeks a theory. Induction seeks for facts. (*EP* 2:106.)

> Abduction is the process of forming an explanatory hypothesis. It is the only logical operation which introduces any new idea; for induction does nothing but determine a value and deduction merely evolves the necessary consequences of a pure hypothesis.
>
> Deduction proves that something *must* be, Induction shows that something *actually is* operative, Abduction merely suggests that something *may be*. Its only justification is that from its suggestion deduction can draw a prediction which can be tested by induction and

[14]With different choices of C and A, this canonical schema may include Schurz's (2008a) "theoretical-model abduction" and "second-order existential abduction". Hoffmann (2010) classifies abductions into five classes, depending on their inference to facts (observable or unobservable), types (concepts), laws, theoretical models, and systems of representation.

[15]Hempel (1965), 246, required that a satisfactory explanation should be *true*, but when the truth value of the explanans is unknown we speak of *potential* explanation.

> that, if we are ever to learn anything or to understand phenomena at all, it must be by abduction that this is to be brought about. (*EP* 2:216.)

Peirce's account here is in harmony with the method of hypothesis, as advocated earlier by William Whewell and later by Karl Popper (see Niiniluoto 1978), but Peirce, like Whewell, admitted that there is an inference from our original problem situation to a hypothesis.[16] After the abductive step in proposing the hypothesis A, the next step is to derive consequences from A by deduction, and then to put theory A into a severe observational or experimental test (*EP* 2:114). This follows the hypothetico-deductive (HD) model of science, but Peirce also allowed for cases where the test evidence is only a probabilistic consequence of the hypothesis.[17] A limiting case of abduction is perception (*CP* 5.181, *EP* 2:227), but abduction may lead backwards to historical causes and to general scientific theories as well. As a scientific realist, Peirce argued sharply against the positivism of Auguste Comte that science should not be restricted to hypotheses "verifiable by direct observation", since such a restriction would "cut off all history as an admissible hypothesis" (*EP* 2:225).

1.4 Interpretations of Abduction

The pragmatist followers of Peirce, like John Dewey, had only a marginal interest in the notion of abduction.[18] The first collection of Peirce's essays, *Chance, Love, and Logic*, edited by Morris Cohen in 1923, included the 1878 article on deduction, induction, and hypothesis, but the classical textbook *An Introduction to Logic and Scientific Method* in 1934 by Cohen and Ernest Nagel only briefly give examples of "presumption of facts" (without reference to Peirce) and a footnote on Peirce's account of Kepler's retroductive reasoning. Even after the first six volumes of Peirce's *Collected Papers* in 1931–1935 and the two additional volumes in 1958, leading philosophers (like Nagel, Reichenbach, and R. B. Braithwaite) were inclined to think that the idea of abductive reasoning as a combination of creativity and inference is in some way confused.

A new way of looking at Peirce's contribution was started by Arthur Burks (1946), who distinguished between Peirce's early treatment of hypothesis as "evidencing" probable reasoning and his "methodeutical" approach after 1891. When new material from unpublished manuscripts became available, commentators like K. T. Fann (1970) have been sensitive to the changes in Peirce's treatment of abduction.[19]

---

[16]Unlike Peirce, Whewell called such inference "induction" (see Sect. 5.1).

[17]Cf. the notion of hypothetico-inductive inference in Niiniluoto and Tuomela (1973).

[18]However, in 1917 F. S. C. Schiller distinguished the logic of discovery from logical proof.

[19]See also Kapitan (1997), Niiniluoto (1999b, 2004), Psillos (2011).

In the 1930s Karl Popper and Hans Reichenbach emphasized the distinction between the context of discovery and the context of justification – arguing that only the latter belongs to the province of logical reconstruction. So it is natural that Peirce's theory of scientific inference was interpreted in the light of this distinction. Burks (1946) suggested that Peirce's way of combining *logica docens* (critical and educational logic) and *logica utens* (native and instinctive habits of reasoning) gives rise to a logic of discovery. With references to the pragmatist F. C. S. Schiller, Norwood Russell Hanson in *Patterns of Discovery* (1958) argued that Peirce's abduction gives a basis for studying the logic of discovery. This proposal was questioned by Harry Frankfurt (1958). Hanson (1961) distinguished reasons for suggesting a hypothesis and reasons for accepting a hypothesis. The new "friends of discovery" started to analyze role of abduction as the heuristic function of the *discovery* of new theories, while an alternative interpretation linked abduction to the motives for investigating or *pursuing* testworthy hypotheses (see Nickles 1980a, b; McKaughan 2008).[20]

The interpretation of abduction as discovery or pursuit restricts its role to the *suggestion* or *generation* of hypotheses for further investigation – such serious hypotheses are potential answers to an explanatory question (see Levi 1979). Such generation may be *selective*, if it tries to find "abducibles" from a pre-existing list of candidates of explanation (see Magnani 2001, 20). This is typical in singular abduction, where the space of possible causes is often very large (Schurz 2008a, 202). For example, medical diagnosis is based on fixed well-established lists of diseases and causes of death. But in theoretical abduction it may be difficult to find even one plausible explanation (see Whewell 1860, 271), so that such inferences are *creative*.[21]

A broader view allows that abduction has a function in the selection and evaluation of explanatory hypotheses which belongs to the context of *justification*. The task of abduction may include the *comparison* of potential explanations and the attempt to find the most promising or *best* among them. Thus, it is not clear that abduction as a method of generation or a motive for pursuit can always be sharply separated from considerations of justification, where the role of abduction in justification can be understood in the weak sense as confirmation and in the strong sense as acceptance (Niiniluoto 1999b; Schurz 2008a).

Peirce allowed abduction to "include a preference for any one hypothesis over others which would equally explain the facts", as long as this preference is not based upon any previous knowledge about the hypothesis, nor on any testing of any of the hypotheses (*CP* 6.525). In his manuscript "The Logic of Drawing History from Ancient Documents" (1901) he formulated principles like explanatory power,

[20]The GW model of abduction (Gabbay and Woods 2005) can be understood as a special case of the pursuit interpretation (see Sect. 5.3).

[21]Hoffmann (2010) distinguishes to kinds of creative hypotheses: new for us but exist in our culture and historically new. The former can be rediscovered, just as Copernicus reconsidered the ancient heliocentric hypothesis.

testability, "caution, breath, and incomplexity" which "guide us in abduction or the process of choosing a hypothesis" (*CP* 7.220, *EP* 2:106–109; cf. Thagard 1981). In addition to these criteria, which belong to the "before-trial evaluation" of a hypothesis (cf. Niiniluoto 1984, 349), the overall assessment of a hypothesis includes the process of testing a hypothesis by empirical evidence.

According to the standard view about the HD method, negative test results disconfirm or refute a general theory, positive results *confirm* the hypothetical theory or give *inductive support* to it (Hempel 1965). Howard Smokler (1968) proposed that "abductive inference" satisfies different principles of qualitative confirmation than enumerative and eliminative induction. Even though Peirce himself as a frequentist was sharply critical of the classical Bayesian theory of probabilistic inference (see e.g. *EP* 2:215), it can be argued that Bayesianism provides a useful framework for studying abduction and induction as forms of ampliative reasoning (see Chap. 6 below). The crucial question is the relation of explanatory power to empirical confirmation. Such study should assess the claim against "explanationism" that only predictive success is relevant in the context of confirmation (see Salmon 2001a).

Peirce himself pointed out that in some cases abductive inference is "compelling" (e.g., perception in ordinary circumstances) or "perfectly certain" (e.g., the existence of Napoleon Bonaparte) (*EP* 2:54). Therefore, it seems appropriate to study the strength of specific abductive arguments such as (3), (6), (9), or (16). By comparison, here we may refer to the enormous literature studying the validity of ampliative inductive arguments of the form (4) or (15). Such studies involve the strong notion of justification as *acceptance*: a hypothesis is so strongly supported by evidence that it is rational to tentatively accept it as true, i.e., to include it by inference in the body of what is currently taken to be scientific knowledge. This kind of proposal was given by Gilbert Harman (1965) with his notion of "inference to the best explanation" (IBE). As an interpretation of abductive reasoning, Harman's IBE claims that the best of rival explanations of given evidence is inductively acceptable on E. Harman also tried to reduce all inductive inference to IBE.

One may question whether the sharp dichotomy between discovery and justification is useful at all, as abduction clearly is related to both of them. Abduction plays an important *strategic* role in truth-seeking (cf. Hintikka 1998; Paavola 2004; Schurz 2008a). From the strategic viewpoint one might suggest that the separate evaluation of specific abductive inferences is unnecessary. Instead, the relevant question is whether such steps contribute to the overall or ultimate goal of inquiry. In this sense, Peirce himself defended abduction as the only form of inference that introduces new ideas. Abduction as such is "nothing but guessing" (*EP* 2:107). But Peirce repeatedly appealed to Galileo's notion of *il lume naturale*, our natural disposition to light upon the correct hypothesis in a finite number of guesses (*CP* 7.223). The justification of the abductively conjectured hypothesis, which is a "may-be" or a possible truth, is then left to the process of estimating its truth-value on the basis of its success in explanation and prediction.

Peirce's later view of abduction, deduction, and induction as successive steps in inquiry should be understood in a dynamic way: after testing abductive conjectures, the cycle starts again (cf. Minnameier 2004).[22] All temporary results of inquiry can be questioned and corrected by further inquiry: what are called "established truths" in science are "propositions into which the economy of endeavor prescribes that for the time being further inquiry shall cease" (*EP* 2:56) Abduction is an essential element of self-corrective science which recursively moves back and forth between observations and generalizations: abduction is a creative inferential process producing theoretical hypotheses on the basis of surprising evidence, and for such hypotheses we find new observations by deduction and induction, which again give rise to new corrected theoretical interpretations.[23] In this sense, we can understand Peirce's praise of the role of abduction in the approach to the truth:

> Abduction is reasoning, which professes to be such, that in the case there is any ascertainable truth concerning the matter in hand, the general method of this reasoning though not necessarily each special application of it must eventually approximate the truth. (Peirce 1976, 37.)

1.5 Some Debates About Abduction

As we have seen, Peirce's own method of analyzing the validity of inferences in the 1870s was based on the idea of truth-frequencies. As deductive inference is necessarily truth-preserving, its truth-frequency is one. For simple probable deduction (13), this value is r, i.e., repeated application of this schema yields a true consequence in the proportion r of all cases. However, the truth-frequency q of the probabilistic hypothetical inference (16) in general is not equal to r, as in (13), since in this inverse inference q depends on the proportion of Fs in the class of Gs, and it is clear that this truth-frequency is independent of r. Indeed, it is possible that it is close to zero. In particular, the truth-frequency of the hypothetical inference (3) is not equal to 1, which follows from the fact that (3) is not deductively valid. In his later work in 1902, Peirce concluded that "probability proper has nothing to do with the validity of Abduction" (*CP* 2.102).

But did Peirce really "mix up" hypothetical reasoning and induction, as he claimed (*CP* 8.227)? After all, the logical structure of arguments (2) and (3) is clearly different. And the context-dependent truth-frequency of (3) may be low, but also high in some situations. Given the rule that all eagles can fly, the inference from "b flies" to "b is an eagle" is very weak (i.e. the proportion of eagles among flying

---

[22]This cycle can be compared to Peirce's schema for the acquisition of beliefs (→ belief → habit → action → surprise → doubt → inquiry → belief →) (*CP* 5.12) and Popper's (1972) schema for the growth of knowledge (problem → tentative theory → error elimination → problem) (see Niiniluoto 1984, 26-27).

[23]Abductive analysis along these lines has been defended by Tavory and Timmermans (2014). Similarly, when McMullin (1992) states that the "retroductive" search of causal explanations is "the inference that makes science", he is referring to the combination of deduction, induction, and abduction.

animals is low). But given the rule that most people living in China speak Chinese, the inference from "b speaks Chinese" to "b lives in China" is strong.[24]

In those cases where the probabilistic validity of the abductive inference (16) is high, Richard Fumerton (1980) has suggested that it could be reduced to a probabilistic argument of the form (13) with the additional premise 'The proportion q of the Gs are F'. Thus, Fumerton proposes to solve the "boundary problem" between abduction and induction (cf. Plutynski 2011) by the claim that abduction can be reduced to inductive arguments: abductions are enthymemes with missing premises. His own example concerns the abductive inference that Jimmy Carter walked on the beach from the premises "there are footprints on the beach" and "if Jimmy Carter had walked on the beach recently, there would be footprints". By adding the missing premises that in all or most cases footprints are imprinted by human beings and that the only person in the beach could have been Jimmy Carter, the abduction is transformed to a valid deduction or induction.

Fumerton's argument about the schema (13) is as such clever, but it can hardly show that all abductions can be reduced to inductions. Rather, it is based on the fact that as ampliative inferences abduction and induction are *non-monotonic* in the sense that adding information to the premises can change the conclusion. The original abduction is defeated, if it is known that Carter was not even close to the beach. In many cases, where an abductive inference has to be made, we are not in the position to add relevant background premises. Similarly, the distinction between deduction and induction does not disappear even though some inductions could be transformed to deductively valid arguments by adding sufficiently strong background assumptions about the uniformity of nature[25]: with the additional premise that all ravens have the same color, one observation of a black raven deductively entails that all ravens are black.

Another reduction is proposed by Gilbert Harman (1968) who argues that inductive generalization is an instance of IBE: from observing instances of F and G, the conclusion that all Fs are G is the best explanation of the premises (cf. (4)). In the DN-model the law that all ravens are black, with the initial condition that this bird is a raven, can be used to explain why this bird is black (cf. (1)). But the law alone is not an explanation of why this bird is a black raven, i.e., general laws do not explain their particular instances.[26] Josephson and Josephson (1994, 18–22) change the explanandum from the instances themselves to the events of observing these instances, and Psillos (2000, 63) to the observed frequencies of Gs among Fs. If

[24]Peirce was well aware of these issues. In 1883 he wrote: "For, in general, there is no connection between the proportion of Ms that are Ps and the proportion of non-Ps that are non-Ms. A very small proportion of calves may be monstrosities, and yet a very large proportion of monstrosities may be calves." (*CP* 2.729, note)

[25]This suggestion was made already by Archbishop Richard Whately in his *Element of Logic* in 1827.

[26]Ennis (1968) gives an example where the inductive generalization is not explanatory (e.g. the correlation between thermometer readings and storm), since its truth is due to common background causes.

inductive generalization is construed as inference from a statistical fact about a random sample to a statistical fact about a population (see (15)), then indeed its conclusion gives a statistical explanation of the form (14). But even if this granted, it would not show that all forms of induction are reducible to IBE: in singular induction or "eduction" (as W. E. Johnson called it), the conclusion is a statement about the next instance to be observed (e.g. the next raven is black), and this cannot serve as the explanation of the premises. Further, the most advanced accounts of inductive generalization analyze induction without appealing to the notion of explanation: in Jaakko Hintikka's system of inductive logic the investigator's willingness to make generalizations on the basis of empirical data depends on contextual regularity assumptions about the universe of discourse (see Niiniluoto 2011c).[27]

While some philosophers use the terms 'abduction' and 'inference to the best explanation' as synonyms (see Thagard 1978; van Fraassen 1989; Kuipers 1999; Walton 2004; Lipton 2004; Psillos 2011; Williamson 2016), many Peirce-scholars have been concerned that abduction should not be confused with IBE. For example, Minnameier (2004) argues that such a confusion would violate Peirce's distinction between abduction and induction. It is true that some advocates of IBE regards it as a species of induction (see Lipton 2004), but still there are types of IBE which allow for the introduction of new explanatory concepts and therefore cannot be formulated as Peircean qualitative induction.

Daniel Campos (2011) argues that abduction should be restricted to the generation of explanatory hypotheses, while IBE involves both the generation and evaluation of hypotheses. He disagrees with Fann (1970), who thought that Peirce did not make a clear distinction between the construction and selection of hypotheses. But when Campos acknowledges that Peirce gave criteria for good hypotheses beyond their power of explanation, the suggested neat division between the generation of plausible conjectures and the evaluation of probable hypotheses becomes problematic.

Especially in later writings, one can find textual evidence that Peirce regarded abduction as an instinct (*CP* 7.220) or guessing (*CP* 7.39) on the basis of non-conscious clues (see Paavola 2005). Still, Peirce maintained till the end that abduction is a form – albeit a weak one – of inference. He found naturalist (evolutionary), idealistic, and theistic grounds for thinking that the first instinctual stage of abduction has some degree of reliability, "strong enough not to be overwhelmingly more often wrong than right" (*CP* 5.173) (see Shanahan 1986). Even though he stated in 1898 that there is "no probability" about abduction, as "no definite probability attaches even to the mode of inference" (*RLT*, 142), he still assumed that abduction has a definite logical form given by (18), which is a natural generalization of his earlier account of hypothesis. We shall see in Chap. 6 that it is possible to associate this canonical schema with Bayesian probabilities which can be updated by new test evidence. This means that the same logical pattern (or its close variations) can serve both discovery and justification in different contexts.

[27]See also Achinstein's (1971) criticism of Harman's argument.

In some of his writings, Peirce made a sharp opposition between "theory", where hypotheses are "tried" and "held provisionally as long as the facts will permit", and "practice" which involves belief and action (*RLT*, 142, 176–78). This thesis that "belief has no place in science" is in conflict with Peirce's famous 1877 characterization of the scientific method in terms of "the fixation of belief" (*CP* 5.358–387). As Benoit Gaultier (2016) observes, this issue is related to the proper formulation Peirce's fallibilism. He further argues that the correct interpretation of the conclusion of abductive reasoning (18), viz. "there is reason to suspect that A is true", is not a recommendation to test whether theory A is true or not but "to determine whether it really is a possible explanation" of the surprising fact C. In my view, this is not plausible, since this proposed conclusion is simply a repetition of the second premise of (18), viz. "But if A were true, C would be a matter of course", which expresses the key connection of abduction with potential explanation. The subsequent task of investigating A is then to determine whether it actually explains C, i.e., whether A is true or not.

When Peirce characterized abduction as an inference to an explanation, he did not give explicit criteria for determining which of the potential explanations is the best one, so that in this sense his most famous schema (18) is different from IBE (Douven 2011). But we have seen that Peirce had interest in the task of selecting the best explanation, so that it is a natural extension of his approach to supply and elaborate such criteria of assessing the power of rival hypothetical explanations (cf. Psillos 2011, 136). In any case, to neglect IBE would give only an incomplete treatment of Peirce's overall account of the self-corrective methodology of science.

Hintikka's (1998) arguments against IBE are based on his strategic view of abduction. But we shall see in Sect. 3.5 that it is possible to incorporate non-deductive abductive inferences into his interrogative model of inquiry.

For these reasons, in this book we shall give a broad survey of different aspects of Peirce's great idea of abductive inference – including its various roles in truth-seeking, discovery, pursuit, strategic reasoning, confirmation, and inference to the best explanation. On many occasions we shall return to Peirce's insights, but – keeping in mind his principles of the "ethics of terminology" – also feel free to develop them in further directions.[28]

[28]For Peirce's 1903 essay "The Ethics of Terminology", see *CP* 2.219–226 or *EP* 2:263–266.

Chapter 2
Analysis and Synthesis

Abstract In spite of his unusually broad knowledge of the history of science, Peirce did not pay attention to the most significant key idea in the history of heuristic reasoning and problem-solving, viz. the method of analysis and synthesis in Greek geometry. As described by Pappus (c. 300 AD), analysis is inverse inference from a theorem to axioms, or from a problem to its solutions, and synthesis then gives the desired direct proof or construction (Sect. 2.2). This inference resembles the regressive method of Renaissance Aristotelians, consisting of a "resolution" from facts to their causes and a "composition" from causes to effects. These methods influenced also such great figures of modern science as Galileo Galilei, Isaac Newton, and Karl Marx (Sect. 2.1). It is argued in this chapter that Peirce's description of hypothesis, as a retroductive inference of a cause from its effect, is an instance of what Jaakko Hintikka calls the upward propositional interpretation of theoretical analysis. Further, the backward solution of a crime case by a detective is an instance of problematic analysis. This thesis is vindicated by Edgar Allan Poe's stories of ratiocination written in the 1840s (Sect. 2.4). Another illustration of the same idea is given in Poe's essay "Philosophy of Composition" where he describes the analytical construction of his poem *The Raven* (1845) (Sect. 2.5).

Charles S. Peirce stated in 1902 that in his study of abduction he was "an explorer upon untrodden ground" (*CP* 2.102; Psillos 2011). It is clear that no one can compete with Peirce as the most original and profound thinker in this theme. But, in spite of his unusually broad knowledge of the history of science, he did not pay attention to the most significant key idea in the history of heuristic reasoning and problem-solving, viz. the method of analysis and synthesis in Greek geometry. It is argued in this chapter that Peirce's description of hypothesis, as a retroductive inference of a cause from its effect, is an instance of what Jaakko Hintikka calls the upward propositional interpretation of theoretical analysis. Further, the backward solution of a crime case by a detective is an instance of problematic analysis. This thesis is vindicated by Edgar Allan Poe's stories of ratiocination written in the 1840s.

I. Niiniluoto, *Truth-Seeking by Abduction*, Synthese Library 400,
https://doi.org/10.1007/978-3-319-99157-3_2

Another illustration of the same idea is given in Poe's essay "Philosophy of Composition" where he describes the analytical construction of his poem *The Raven* (1845).[1]

2.1 The Regressive Method

Peirce's early account of hypothetical reasoning as the inference of the minor premise of a syllogism, or the initial condition of an explanation, resembles Aristotle's examples of "quick wit", defined as the power of "hitting upon the middle term without a moment's hesitation" (*An. Post.* I, 34). For example, we see someone talking to a rich man, and immediately decide that it is because he is trying to borrow money.

Aristotle's treatment of *epagoge* in *Analytica Priora* (II, 23) involves the inversion of syllogisms, but for him this process in inseparable from concept formation. William Whewell, in *Philosophy of Discovery* (1860), followed Aristotle closely, as in his "inductions" there is always "a new conception", suggested by "the sagacity of discoverers" (see Niiniluoto 1978, 1995). Peirce, instead, argued in 1883 that in a valid induction the attribute to be detected in the sample has to be "predesignated" or "agreed upon in advance", so that the discovery of laws in science is accomplished by induction, and the discovery of causes by hypothetic inference (*CP* 2.736).

In his 1898 Cambridge Lectures, Peirce suggested that induction corresponds to Aristotle's *epagoge* and abduction to Aristotle's *apagoge* (see *RLT*). The latter term is often translated as "reduction". A historically important related argument is known as *reductio ad absurdum*: from a given statement consequences are derived, and if a false consequence is reached, the original statement must be false as well. But if the new facts are always found to match the derived consequences, the hypothesis has been tested. This kind of inference was called "induction" by Peirceafter 1901.

Peirce's characterization of deductive explanation as reasoning from causes to effects has its roots in Aristotle's doctrine in *Analytica Posteriora* (see I, 13, 34) that a "scientific syllogism" proceeds from causes to their effects. The medieval

[1]This chapter is largely (except Sect. 2.3) based on a paper presented in the conference on Model-Based Reasoning in Scientific Discovery, Pavia, December 17–19, 1998 (see Niiniluoto 1999c). The prehistory of this paper goes back to the year 1975 when I acted as one of the opponents of Unto Remes's (1942–1975) doctoral dissertation on the geometrical method of analysis and synthesis (see Hintikka and Remes 1974). During the disputation, I suggested that Edgar Allan Poe's detective stories follow the method of analysis which has also a close resemblance to Peirce's account of abductive inference. Two years later I found by accident a short article, published by my father Yrjö Niiniluoto (1900–1961) as a young man in 1925, which referred to Poe's fascinating essay on the "philosophy of composition". My paper was published in Finnish in 1978, and later with some revisions in 1990. In the meantime, Thomas A. Sebeok had published in 1980 his article on Peirce and Sherlock Holmes, followed by a collection of essays on abduction and detective stories (see Eco and Sebeok 1983). Therefore, I did not hurry to write an English version of my paper, but waited for a suitable occasion for its presentation until 1998.

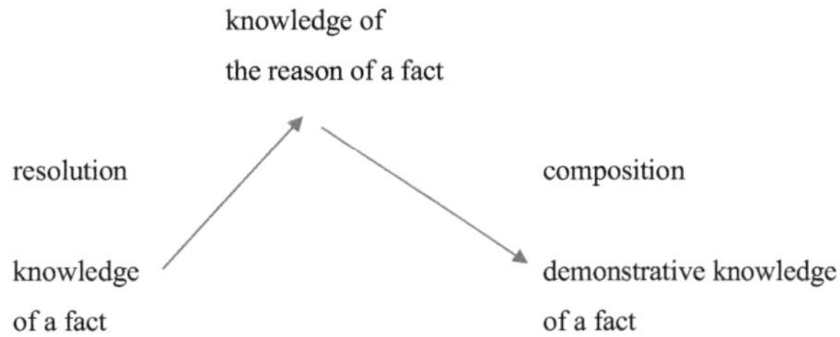

Fig. 2.1 The regressive method

followers of Aristotle and Galenos, from Ali ibn Ridwan (Hali Rodowan) in the eleventh century and Robert Grosseteste in the thirteenth century to Jacopo Zabarella in the sixteenth century, called this inference *demonstratio propter quid* or *compositio*. The inverse inference from effects to their causes was called *demonstratio quia* or *resolutio* (see Crombie 1953). Taking the clue from Aristotle's distinction between "knowledge of a fact" and "knowledge of the reason of a fact" (*An. Post.* II, 13), the mode of presenting science was assumed to follow a "regressive" method (see McMullin 1992): starting from something that is observable, better known to us, an effect, complex, and singular, we proceed to something which is better known to nature, a cause, simple, and general, and then back to the original phenomenon, but now "knowledge-that" replaced by "knowledge-why" (see Fig. 2.1).[2] Some Aristotelians believed that resolution can proceed by deductively valid syllogisms, some identified this inference with induction.

The regressive method can be illustrated by medical diagnosis: resolution leads from symptoms to a disease, and composition back from the disease to the symptoms. If a person has fever, it may have many alternative causes, but a more detailed description of the symptoms may justify inference to a unique disease. If it is known as a rule that "A person has measles if and only if she has fever, spots in the mouth, and red rash in the face and body", then one can construe a deductive syllogism from these symptoms to the measles. This resolution from effects to their cause serves discovery or diagnosis. Then composition gives demonstrative knowledge of the symptoms or explains them by a scientific syllogism from causes to effects.

The main idea of the regressive method, including the terminology with "resolutions" and "compositions" was followed by such great figures of modern science as Galileo Galilei, Isaac Newton, and Karl Marx (see Burtt 1952; Hintikka and Remes 1974). Galileo combined this method to the mathematical description of nature. Newton argued in his *Opticks* (1704) that "as in Mathematics, so in Natural Philosophy, the Investigation of difficult Things by the method of Analysis, ought ever to precede the Method of Composition", where such Analysis proceeds "from Compounds to Ingredients, and from Motions to the Forces producing them; and in general, from Effects to their Causes". For Marx, who followed Hegel's terminology, analysis reveals the "essence" of economic phenomena behind visible

[2] As far as I know, Peirce never treated the regressive method, but in 1898 he remarked that Aristotle "was driven to his strange distinction between what is better known to Nature and what is better known to us" (*EP* 2:43).

appearances, and synthesis "rises from the abstract to the concrete". But in modern science, these procedures were regarded as modes of inquiry leading to new knowledge – rather than as modes of teaching and organizing old knowledge.

In this light, the novelty in Peirce's theory of abduction is not in the claim that there is a type of inference from effects to their causes. Peirce's originality can be seen in his attempt to generalize Aristotelian logic, including the statistical and probabilistic forms of inference.

Another interesting aspect of Peirce is related to his dynamic conception of laws. For Aristotle, the laws of nature expressed essential connections between qualitative properties, and this was true of his talk about causes and effects as well. But the modern conception takes causal laws to be dynamical or relational invariances[3] – "laws of succession" in J. S. Mill's sense. Thus, one way of understanding the general rule of a syllogism is that it expresses a causal law. As causes temporally precede their effects, the inverse inference from effects to their causes involves reasoning backwards in time. If we know that lightning is always causally followed by thunder, then from a thunder we can reason backward to the event of lighting (even its distance from us). This temporal interpretation of hypothetic inference motivates Peirce's term *retroduction* and some of his examples: perceptual judgments as "an extreme case of abductive inferences" (*CP* 5.181–185) and our knowledge of history (*CP* 2.714).

2.2 Pappus on Geometrical Analysis

The Latin terms *resolutio* and *compositio* are translations of the Greek terms *analysis* and *synthesis*. Against J. H. Randall's (1961) thesis that Galileo and Newton adopted the method of science from the regressive method of the Paduan Aristotelians, it has been argued that their main inspiration should be seen to arise from the method of analysis and synthesis of Greek geometry (see Hintikka and Remes 1974)The most famous description of this "secret" of the Greek geometers, or the "Treasury of Analysis", was given by Pappus (ca. 300 AD), whose work was translated into Latin in 1566, and was known to the great champions of the new mathematical physics.

The method of analysis and synthesis was known already in Plato's school. In *Nicomachean Ethics* (1112b15-29), Aristotle compares it to the structure of deliberation: in the process of planning, a decision-maker searches for means to obtain given ends, and further means to obtain intermediate means, etc., until this reasoning "backward" comes to something that can be done or is impossible. In the former

[3]The distinction between Aristotle's substantial invariances and Galileo's dynamical invariances was made by Eino Kaila in 1939 (see Kaila 2014).

case, the synthesis reverses the steps in order to reach the desired goal. Hence, analysis goes here from ends to means, and synthesis conversely from means to ends.[4]

The famous description by Pappus is as follows[5]:

> Now analysis is the path from what one is seeking, as if it were established, by way of its consequences, to something that is established by synthesis. That is to say, in analysis we assume what is sought as if it had been achieved, and look for the thing from which it follows, and again what comes before that, until by regressing in this way we come upon some one of the things that are already known, or that occupy the rank of a first principle. We call this method 'analysis', as if to say 'reduction backward'. In synthesis, by reversal, we assume what was obtained last in analysis to have been achieved already, and setting now in natural order, as precedents, what before were following, and fitting to each other, we attain the end of the construction of what was sought. This is what we call 'synthesis'.

Pappus goes on to explain that analysis is two kinds: "theoretical" which seeks after truth, and "problematic" which tries to find or construct what was demanded.

> In the case of the theoretical kind, we assume what is sought as a fact and true, then, advancing through its consequences, as if they are true facts according to the hypothesis, to something established; if this thing that has been established is a truth, then that which was sought will also be true, and its proof the reverse of the analysis; but if we should meet something established to be false, then the thing that was sought too will be false. In the case of the problematic kind, we assume the proposition as something we know, then, proceeding through its consequences, as if true, to something established; if the established thing is possible and obtainable, which is what the mathematicians call 'given', the required thing will also be possible, and again the proof will be reverse of analysis; but should we meet with something established to be impossible, then the problem too will be impossible. Diorism is the preliminary distinction of when, how, and in how many ways the problem will be possible.

There is a long-standing and lively debate on the meaning of this passage and its interpretation (see, e.g., Heath 1956, vol. 1, 137–142; Hintikka and Remes 1974; Mäenpää 1997). The "standard" interpretation of theoretical analysis is as follows: from a proposed theorem T_0 a logical consequence T_1 is deduced, from that T_2, etc. until a statement T_n known to be false or true is reached; in the former case, T_0 is false as well, in the latter case, T_0 can be proved in synthesis from T_n, provided all the previous deductive steps from T_i to T_{i+1} are reversible, i.e., T_i and T_{i+1} are in fact logically equivalent.[6] This is called "downward propositional interpretation" of analysis by Hintikka and Remes (1974). The "upward propositional interpretation", instead, suggests that analysis seeks premises from which T_0 can be deduced, so that it is always required that T_{i+1} entails T_i for all $i = 0, 1, \ldots, n\text{-}1$.

We can now see that these propositional interpretations make the method of analysis and synthesis resemble closely the regressive method: analysis reasons

[4]Hilpinen (2007) argues that practical reasoning, following Aristotelean practical syllogisms, is a form of abduction in Peirce's sense.

[5]See Pappus (1986), 82.

[6]See the essay by A. Szabó in Hintikka and Remes (1974), 118–130.

backwards from a hypothetical theorem (T_0) to some axioms of geometry (T_n), and synthesis gives the proof of T_0 as a theorem on the basis of T_n.[7]

We also see that Peirce's description (1.18) of abduction corresponds to the upward propositional interpretation of analysis: a surprising fact is given, and a theory from which it follows (by deduction or probability) is sought. However, in the case of natural science, we do not have an independent or direct way of verifying the theory, and there may several rival theories which would all explain the given fact. Therefore, abduction only leads to potential theoretical explanations which are testworthy, and to choose between them further "inductive" tests by their other consequences have to be performed.

Hintikka and Remes (1974) make important objections to the propositional interpretations of analysis. One of their observations is that theorems in geometry are typically general statements (e.g., 'for all triangles, the sum of their angles equal 180°') or universal-existential statements (e.g., 'for all geometrical figures x, if x is a square, then x will have two diagonals and these diagonals bisect each other'). A proof of such general implications proceeds through their instantiations and by attempting to derive the consequent from the antecedent by suitable axioms or rules of inference. As worked out to some extent by Hintikka and Remes, one particularly effective way of attempting to construct such proofs is Beth's method of semantic tableaux. E.W. Beth himself suggested that his semantic tables might be understood as an explication of the method of analysis. This in a sense reconfirms the intimate connection between analysis and Peircean abduction.[8]

Another observation by Hintikka and Remes is that a proof of general implications in ordinary predicate logic may need the use of instantiation rules by which new individuals are introduced into the inference. Hintikka has used the amount of such needed additional individuals as a measure of "non-triviality" of deductive inference. Again, Peirce anticipated this idea in his discussion of trivial or "corollarial" deductive reasoning in contrast to "theorematic" ones (*CP* 2.267; see Hintikka 1980). In geometrical proofs, such new individuals correspond to auxiliary constructions that are drawn in figures and illustrations. In this view, the heuristic power of analysis lies precisely in these auxiliary constructions: as first-order logic is not effectively decidable, there are no mechanical methods of making such constructions or even knowing their number in a particular case. Analysis is not a "foolproof" algorithm for discovery, but as soon as the necessary constructions

[7]It is ironic that the custos (chairman) of the doctoral disputation of Unto Remes (see note 1) was Oiva Ketonen, who in his 1944 doctoral thesis had improved Gentzen's sequent calculus into a system with invertible rules: if the conclusion of a rule follows from the premises, the premises are derivable from the conclusion. This system was generalized from propositional logic to predicate logic by S. C. Kleene in 1952. In an invertible system the distinction between downward and upward analysis disappears. Ketonen never mentioned this conclusion in our discussions, but a sophisticated treatment of his system with applications to automated deduction is given by Sara Negri and Jan von Plato (2001).

[8]The method of Beth tableaux has been used by Atocha Aliseda (1997) in her treatment of abduction. See Sect. 3.3.

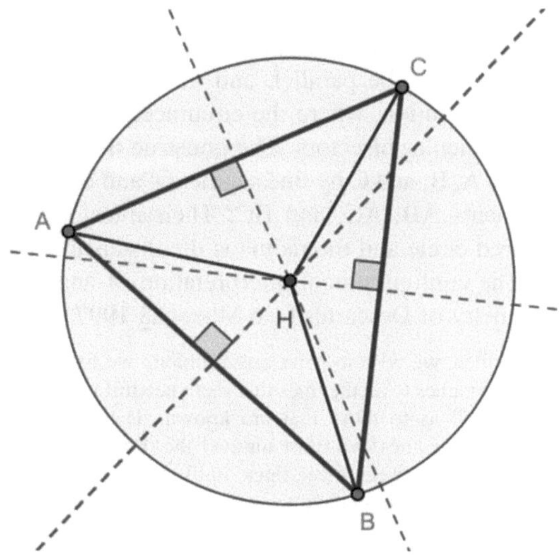

Fig 2.2 Model figure of a problematic analysis

have been found by the wit or imagination of the researcher, it specifies steps for making genuine discoveries.

These considerations led Hintikka and Remes to replace the propositional interpretation of analysis by what they call the "analysis of figures" or "configurational analysis". Instead of being a method of seeking deductive connections between propositions, analysis and synthesis consider geometrical figures and the functional interconnections between their elements. Following Kant, Peirce also argued that such "observation and experiment" of "a diagram of our own creation" is an essential feature of all "necessary reasoning" in geometry (*CP* 3.554–560). The configurational idea is especially clear in the case of problematic analysis which attempts to solve geometrical problems.[9]

In twentieth century textbooks of school mathematics, the method of solving geometrical problems by analysis and synthesis was sometimes called "the method of model figures". The solution of a problem starts by assuming that the desired construction has already been accomplished. Then some auxiliary constructions are added to this model picture and the relations of its parts are studied. Eventually a condition for the thing sought is obtained in this "backward solution", and the number of ways of realizing this condition is counted as the number of possible solutions.

Let us illustrate this method by the problem of drawing a circle through three given points A, B, and C (see Fig. 2.2). To derive the solution by analysis, assume that the desired circle with center point H has been drawn. Join the points A, B, and C by three line segments. Then H must be equally distant from A and B, so that it is

[9]Petri Mäenpää's (1997) elegant formalization of this method uses, instead of predicate logic, Per Martin-Löf's type theory to account for constructions and their results.

situated in the perpendicular bisector of the line segment AB, and similarly for the line segments AC and BC. If the points A, B, and C are on the same line, these bisectors will be parallel, and the problem has no solutions. Otherwise there is a unique solution, where the circumcenter H is the point of intersection of the three perpendicular bisectors. The construe the solution in synthesis, start by joining the points A, B, and C by line segments, and draw the perpendicular bisectors to the line segments AB, AC, and BC. Their intersection is then the center point H of the desired circle and the radius is the distance between H and A.

The configurational interpretation of analysis was also the basis of the analytic geometry of Descartes (see Mäenpää 1997):

> If, then, we wish to solve any problem, we first suppose the solution already effected, and give names to all the lines that seem needful for its construction, – to those that are unknown as well as to those that are known. Then, making no distinction between known and unknown lines, we must unravel the difficulty in any way that shows most naturally the relations between these lines, until we find it possible to express a single quantity in two ways. This will constitute an equation ... (Descartes 1954, 6.)

So the trick is again to draw a model figure, maybe with additional constructions, identify the known elements, and mark the unknown as 'x'. Then, treating x as if it were known, an equation is established, and finally the equation is solved with respect to x.

2.3 Analysis and Abduction

Jaakko Hintikka's work on geometrical analysis was inspired by his proof-theoretical studies in the 1950s. His distributive normal forms gave a natural way of distinguishing trival and non-trivial deductions in first-order logic, where the latter increase the depth of formulas, measured by the number of considered objects insofar as they are introduced by individual constants or layers of quantifiers. He applied this technique to Kant's philosophy of mathematics, which emphasized the need of auxiliary constructions in proofs, with the conclusion that logical and geometrical truths are synthetic a priori (see Hintikka 1973). It was only later that Hintikka, to his pleasure, found out that Peirce had made a similar distinction between "corollarial" and "theorematic" deductions (see Hintikka 1980), where in the former it is "only necessary to imagine any case in which the premises are true in order to perceive immediately that the conclusion holds in that case", while in the latter it is "necessary to experiment in imagination upon the image of the premiss" (see Peirce 1976, vol. 4, 38) by introducing new or "foreign ideas" to it (*ibid.*, 42).

Hintikka and Remes (1974), 6, mention Peirce's logical distinction only in a footnote, and they do not relate the method of analysis and synthesis to Peirce's notion of abduction. But we have, following Niiniluoto (1999c), argued that theoretical and problematic analysis are abductive in the sense that they proceed backward from theorems to axioms or from desired constructions to their conditions. More formally, the upward propositional interpretation analysis (like abduction) seeks premises from which the theorem can be derived. Still, a qualification may be added: such analytical inferences are not triggered by "surprising facts" and the produced synthesis is a proof or problem solution rather than an explanation. In this respect the scientific counterparts of geometrical analysis and synthesis, viz. resolution and composition of the regressive method, are even closer to Peirce's characterization of abduction.[10] In the next sections, we shall see that this is also the case with analytical inferences in detective stories.

Ketner (1985) has argued that Hintikka misunderstood Peirce's corollarial – theorematic distinction, but this is unfair, as Webb's (2006) discussion and assessment shows. Ketner's own treatment allows very simply inferences in propositional logic to be theorematic, which misses Peirce's intention to capture "the great difference between corollaries and major theorems" in mathematics (Peirce 1976, vol. 4, 1). Peirce had keen interest in the question whether machines can perform mathematical deductions (see Webb 2006), and – even without knowledge of the later undecidability results about logic – he understood that theorematic deductions are creative in an important sense. In 1907 he even coined a new term 'theoric deduction' to parts of demonstrations with "the power of looking at facts from a novel point of view", so that they, even though "indisputable", are "very plainly allied to retroduction" (see Hoffmann 2010).

In spite of the creative element in theorematic reasoning, one should not give up the conceptual distinction between deduction and abduction. As Hintikka and Remes (1974), 4, correctly emphasize, in their configurational reading both analysis and synthesis make use of auxiliary constructions, which is confirmed by our example with Fig. 2.2. Hence, both analysis and synthesis are typically instances of theorematic reasoning, but only analysis has an abductive heuristic character whereas synthesis consists of a proof. There is some confusion about this matter in recent literature.

Ketner (1985), 411, suggests that the production of "experiments" in geometrical and algebraic diagrams is "done through abduction". But e.g. the addition of new points and lines in Fig. 2.2 as such hardly satisfies any of Peirce's conditions of abductive reasoning. Minnameier (2017), 186, proposes that theorematic deduction should be understood as inference "from the results of corollarial deduction to the premises from which the result can be deductively derived".[11] Even though this kind of inverse inference has affinity to abduction, Minnamaier overlooks the fact that theorematic deduction has its prime examples in direct geometrical proofs. Magnani (2009) introduces a distinction between theoretical and manipulative abduction, where the prime example of the latter is diagrammatic reasoning in geometry.[12] Park (2017a), 212, welcomes this proposal, as "the possibility of expanding the

[10]McMullin (1992) also points out that Peirce's abduction is not demonstrative in the same strong sense as resolution was often described in the regressive method.

[11]Minnameier (2017), 188, also proposes a model of "inverse abduction" which seeks concrete instances of abstract theories or concepts.

[12]For some remarks about "manipulative abduction" as a form of practical reasoning, see Sect. 5.3.

scope of abduction is wide open" (see also Park 2017b). But even though the direct demonstration that the sum of angles of any triangle is 180° uses "manipulation" in the form of auxiliary construction of new lines and angles (see Magnani 2009, 176; Park 2017a, 202), there is no good reason for saying that this direct mathematical proof is "abductive".

2.4 Poe's Stories of Ratiocination

Thomas A. Sebeok and Jean Umiker-Sebeok (1980) tell a true story about an incident in 1879, when Charles Peirce was able to guess the person who stole his watch and overcoat on a boat trip. They argue that this inference, which was later verified by a house search, was an example of Peircean abduction.

Sir Arthur Conan Doyle (1859–1930) published his first Sherlock Holmes story, *A Study in Scarlet* in 1888 (see Doyle 1952). There is no direct evidence that Peirce had read Holmes stories, but Edgar Allan Poe (1809–1849) seems to belong to his favorite writers.[13] Poe published his first detective story *The Murders in the Rue Morgue* in 1841, followed by *The Mystery of Marie Rogêt* in 1843 and *The Purloined Letter* in 1845 (see Poe 1977). These "stories of ratiocination", with their hero C. Auguste Dupin, mark the birth of the new detective genre in the world literature.

Dupin and Holmes are master detectives who solve crime cases with their powerful methods of reasoning. Carefully observed facts about the case serve as clues for further investigations and questions, but when the necessary information has been collected, the solution is obtained by an intellectual operation. Agatha Christie's Hercule Poirot follows this genre, while Raymond Chandler's Philip Marlowe operates with straightforward action and intervention. Indeed, Chandler's "The Simple Art of Murder" (1964) questions the whole tradition of "logic-and-deduction-novels" with its artificial treatment of "formal and exact problems".

Reasoning in classical detective stories is usually called "deduction". There is no doubt that deduction in the strict sense of logically valid inference has an important role in them. But it is also easy to see that many of the inferential steps and "guesses" that Dupin and Holmes perform during their stage of investigation are abductive or retroductive in Peirce's sense, i.e., they infer a cause from an effect.[14] For example, Holmes notes that a person has in his right arm heavier and bigger muscles than in his left arm and infers that he is a manual worker. When Dr. Watson was introduced to Holmes, the detective immediately concludes that "You have been in Afghani-

[13]Rubén Sampieri Cábal has suggested that Poe's "prose poem" *Eureka* (in 1848) anticipates the ideas of Peirce's abduction.

[14]This is observed also by Sebeok and Umiker-Sebeok (1980); Truzzi (1983), 69; Harrowitz (1983), 185–194.

stan, I perceive", by inferring that Watson has been an army doctor from the dark tint of his face, fair wrists, and injured arm.[15]

More generally, the solution of a murder is a paradigmatic example of retroduction. Given facts about the present that are traces of earlier events, the task of the detective is to give a historical reconstruction of the historical process that led to the death of the victim. It is no surprise that inferential steps leading to this reconstruction are abductive in the Peircean sense.

It is also important to observe that this kind of retroduction involves "solution backward" in the classical geometrical sense of analysis. This is not very much emphasized by Doyle, who himself was educated in medicine and patterned the character and skill of Sherlock Holmes after his professor Joseph Bell of the Royal Infirmary of Edinburgh (see Sebeok and Umiker-Sebeok 1980, 54). There are two well-known passages where Holmes describes his method in terms that are linked to the method of analysis (cf. *ibid.* p. 64, 66), but we shall see that they are directly influenced by Poe. In *A Study in Scarlet*, Holmes states:

> In solving a problem of this sort, the grand thing is to be able to reason backward. That is a very useful accomplishment, and a very easy one, but people do not practice it much. In the everyday affairs of life it is more useful to reason forward, and so the other comes to be neglected. There are fifty who can reason synthetically for one who can reason analytically... There are few people, however, who, if you told them a result, would be able to evolve from their own inner consciousness what the steps were which led up to that result. This power is what I mean when I talk of reasoning backward, or analytically.

In *The Five Orange Pip*, Holmes refers to Cuvier's study of a whole animal "by the contemplation of a single bone", and concludes that

> the ideal reasoner ... would, when he had once been shown a single fact in all its bearings, deduce from it not only the chain of events which led up to it but also the results which would follow from it.

It is remarkable that Edgar Allan Poe was aware that "the analytic power" of Monsieur C. Auguste Dupin is historically associated with the mathematical tradition. Poe, who had studied at the University of Virginia, had a wide interest in the scientific and philosophical literature of his time, as vividly illustrated by his science fiction stories (see Poe 1976). This interest covered also mathematics and its applications, such as cryptography and chess-players (see Poe 1977).

In the opening paragraph of *The Murders in the Rue Morgue*, Poe discusses "the mental features discoursed of as the analytical" (Poe 1977, 332–333):

> The faculty of re-solution is possibly much invigorated by mathematical study, and especially by that highest branch of it which, unjustly, and merely on account of its retrograde operations, has been called, as if *par excellence*, analysis.

---

[15]This example is discussed in Hintikka and Bachman (1991). Hintikka relates the "logic of Sherlock Holmes" to his interrogative model of inquiry, when all inferential steps are deductive while abduction has a role in information-seeking through questioning (see Hintikka and Hintikka 1983; Hintikka 1998). This model of inquiry is discussed in Sect. 3.5.

Poe is referring here to the basic area of modern mathematics, Differential Calculus, known as "the Analysis" and "the Calculus". He objects immediately that "to calculate is not in itself to analyse". Calculation, such as chess-playing, is not yet analysis, since the analytic ability requires imagination which goes beyond mere ingenuity (*ibid.*, 336). In *The Purloined Letter*, mathematical reasoning is characterized as "merely logic applied to observation upon form and quantity" (*ibid.*, p. 453). The criminal Minister in the story is described by Dupin as being both a poet and a mathematician, and therefore he could reason well (*ibid.*, 453). Similarly, Dupin himself is both creative and resolvent. Dupin's criticism of the French "deception" of speaking about "analysis" in connection with algebra suggests, however, that Poe was not acquainted with the historical connection between geometrical analysis and analytical geometry.

In the story about Rue Morgue, two women are found murdered in an unusually brutal way: one in a locked room, strangled and thrown in the chimney, the other in a yard in the rear of the building with her throat cut. The men making these horrible findings hurried to the place alarmed by terrible shrieks from the house. When they rushed to the stairs, they heard an angry contention by two voices, one by a Frenchman and the other by a foreigner with a shrill voice. The murders seemed to have no motive, and the police found no clues for solving them.

After investigating the scene of crime and the testimonies of the witnesses, Dupin solves the mystery by an astounding chain of inference. After excluding the possibility of suicide or supernatural explanations, he concludes that the murders are committed by a third person. But how could the assassin escape from the room? "Fortunately, there is but one mode of reasoning upon the point, and that mode *must* lead us to a define decision", Dupin explains. As other means of regress are excluded, the murderer must have passed through the windows of the back room. In showing that this "apparent impossibility" is a real possibility, Dupin concludes *a posteriori* that there must be something wrong in the nail which fastened the sashes of the window.

Dupin then puts together the "very unusual" strength and agility, brutal cruelty, startling absence of motive, peculiar shrill and unequal voice "about whose nationality no two persons could be found to agree and in whose utterance no syllabification could be detected". With the additional clues of a tuft of hair in the hand of one victim and the facsimile of the finger nails in the throat of the other victim, Dupin infers – by appealing to Cuvier's zoological study – that the murder has been accomplished by an Ourang-Outang of the East Indian Islands.

The structure of Dupin's inference resembles the backward solution of a mathematical problem by means of the method of analysis. His starting point is a situation with a murder (cf. the geometrical model figure), with the so far unidentified murderer (cf. the sought, unknown entity x). The question is how this situation has been generated (cf. the needed constructions for creating the unknown x). Using effectively the information about the situation he develops the solution step by step backward (cf. the retroductive direction of analysis). Progress often requires additional questions and observations (cf. the auxiliary constructions in the model figure), which also serve to eliminate some of the alternative explanations. After the mystery has been solved, the succession of events is reconstructed in its temporal order (cf. the synthesis after an analysis).[16]

The story about Rue Morgue has been designed so that there is only one reasonable, though imaginative solution which explains all the details about the case. It thus corresponds to a geometrical problem with a unique solution. In such a situation the backward analysis can proceed only in one way. As Dupin states about his "legitimate deductions": "the deductions are the *sole* proper ones, and ... the suspicion arises *inevitably* from them as the single result". Still, this analysis requires creative imagination, as the analyst has to know what to ask, observe, and search for.

Poe also lets his hero to verify the inferred conclusion by putting in *Le Monde* an advertisement, telling that "a very large Ourang-Outang of the Bornese species" has been caught in a park and its owner may have it again by paying a few charges. This kind of action might be compared to experiments or manipulations of nature that have an important role in Newton's application of the method of analysis in science.

Poe knew that all problem-solving cannot be so straightforward or compelling as in the story of Rue Morgue, since there may exist alternative backward paths and the task of the investigator is to find the most probable of them. This topic – which of course was crucial in Peirce's probabilistic formulations of abduction as well – is discussed in *The Mystery of Marie Rogêt*. Poe's understanding of the Calculus of Probabilities is not entirely adequate, since he seems to believe in the phenomenon of the "maturity of chances": the occurrence of two sixes in succession with a dice is for him a sufficient cause for betting that sixes will not be thrown in the third attempt (Poe 1977, 438). More interestingly, Poe observes that the probability of a hypothesis is increased in the ratio of the "peculiarity" or "unusualness" of the facts it explains (cf. *CP* 7.182). Such corroborative evidence does not increase probability additively but geometrically: each new proof or piece of evidence multiplies the earlier probability (*ibid.*, 405–406). These ideas could be easily formalized by Bayes's Theorem: if hypothesis H entails evidence E, then the posterior probability P(H/E) of H given E equals the prior probability P(H) of H multiplied by the factor 1/P(E) (see Sect. 6.2).

Poe's story *The Gold-Bug* (1843) illustrates the principle that the most probable explanation deserves to be investigated in a heuristic search. The topic of the story, secret writing or cryptography, was discussed by Poe also in an essay two years earlier (see Poe 1977, 538–546). In the story, William Legrand is able to decode a letter of the pirate Captain Kidd by constructing the key on the basis of the statistical frequencies of letters in ordinary English texts. This method could be presented in any modern textbook of information theory.[17]

[16]Similar analytic or abductive methods of reasoning (like fingerprints and DNA sequences) are used in contemporary forensic science, familiar from the tv-series C.S.I.

[17]Cf. Smith (1955). Beaulieu (2008) gives an interesting account Peirce's contribution to American cryptography, but Poe is not mentioned.

2.5 Poe's Philosophy of Composition

Poe's fascination with mathematical analysis extends to his work as a poet. The construction of works of art requires, besides ingenuity, "the faculty of analysis" (Poe 1927, 326). As Poe remarks on Longfellow, someone may sometimes write brilliant poems "by accident" (*ibid.*, 269). But the proper way of composing poetry must be based upon knowledge of the aims of poesy – which is the creation of beauty rather than truth (*ibid.*, 271, 279) – and of its general principles. The poetry of words is defined by Poe as "the rhythmical creation of beauty" (*ibid.*, 99), and two-thirds of the principles of versification belong to the mathematics (*ibid.*, 333).

Poe's essays "The Poetic Principle", "The Rationale of Verse", "Fancy and Imagination", and "Rhyme" contain many interesting observations. Anticipating modern information-theoretical accounts of aesthetics, he remarks that "perfection of rhyme is attainable only in the combination of the two elements, Equality and Unequality", and "unexpectedness must arise from expectedness" (*ibid.*, 311). Anticipating later doctrines of emergent qualities, Poe points out "the range of Imagination is unlimited", as the admixture of two elements may result in something that has no qualities of either (*ibid.*, 316). And anticipating modern semiotic studies, Poe points out that in a poetic text there may lie "beneath the transparent upper-current of meaning, an under or suggestive one" (*ibid.*, 285).

The essay "The Philosophy of Composition" gives a surprising description of the birth of Poe's perhaps best-known poem *The Raven* (1845):

> It is my design to render it manifest that no one point in its composition is referable either to accident or intuition – that the work proceeded, step by step, to its completion with the precision and rigid consequence of a mathematical problem. (*Ibid.*, 165–166.)

That Poe's account here corresponds to the problematic analysis in the geometrical sense becomes evident when he states that his poem has its beginning "at the end, where all works of art should begin" (*ibid.*, 171). The poet should start by selecting some desired effect, and then find aids in "the construction of the effect" (*ibid.*, 164). The artistic means are then chosen so that "in the whole composition there should be no word written, of which the tendency, direct or indirect, is not to the one pre-established design" (*ibid.*, 188). "It is only with the *dénouement* constantly in view that we can give the plot its indispensable air of consequence..." (*ibid.*, 164).

Poe starts from the intention of composing a poem that suits "at once the popular and the critical taste". From the principles that a poem should not be "too long to be read at one sitting", that beauty is "the sole legitimate province of the poem", and that melancholy is "the most legitimate of all the poetical tones", Poe concludes that the length of the intended poem should be about one hundred lines and it should deal with beauty with the tone of sadness. Using "ordinary induction" Poe then chooses refrain as his artistic effect. The best refrain, he decides, is a single word forming the close of each stanza. It should include the long *o* as the most sonorous vowel and *r* as the most producible consonant. The first word, expressing the melancholy tone, which occurred to Poe is 'Nevermore'. But what would the occasion for the repeated use of the word 'nevermore'? As a rational human being would not fit the purpose, and a parrot would not be in keeping with the intended tone, a raven as "the bird of ill omen" was chosen.

According to Poe, death is most melancholy of all melancholy topics, and the death of a beautiful woman is "unquestionably the most poetical topic in the world". The only intelligible way of combining these ideas is to imagine the raven employing the word 'nevermore' in answer to the queries of a lover. The last of these self-torturing questions should involve "the utmost conceivable amount of sorrow and despair". At this moment Poe puts his pen to paper and writes the climax of the poem:

"Prophet", said I, "thing of evil! Prophet still if bird or devil!
By that heaven that bends above us – by that God we both adore,
Tell this soul with sorrow laden, if within the distant Aidenn,
It shall clasp a sainted maiden whom the angels name Lenore -
Clasp a rare and radiant maiden whom the angels name Lenore."
Quoth the raven "Nevermore."

The writing of the poem thus starts from the end, more precisely, from the sixteenth stanza. After writing the first fifteen stanzas approaching gradually the *dénouement*, where the raven destroys the lover's final demand to meet his mistress in another world, Poe still adds two stanzas at the end to create "suggestive meaning" by metaphorical expressions.

According to Poe's "philosophy of composition", the construction of a poem follows an equally definite pattern as Dupin's inferences in solving crime mysteries. Both are "analytic" in the traditional sense that they start from the end and seek the solution by backward steps. In this sense, Poe generalizes to the realm of poetry and artistic creation the principles of analysis and synthesis by Isaac Newton. Peirce once remarked that physics owes all its triumphs to the "Analytic Method" (*CP* 1.64). Essentially the same method has been applied in engineering design (where construction problems are analyzed into subproblems until we reach something that can be done), and in computer science for the construction of programs (cf. Mäenpää 1997). Analysis was advertised as a general method of problem-solving already by George Polya in his *How to Solve It* in 1945 (see Polya 1957).

Not all commentators have taken seriously Poe's account of his poetic construction "with the precision and rigid consequence of a mathematical problem", since it seems to clash with established views about creativity in art. A hallmark of creation is its non-purposiveness, i.e., "its outcome can be a surprise even to its originator" (Hintikka 1975, 196). But it seems to me that this aspect of creativity can find its place also in Poe's method of composition.

The first part of Poe's description consists of the discovery of the climax of the poem. This stage involves artistic inspiration and imagination, and – in spite of some inferential steps concerning the kind of effect that is desired – it is not yet present within the initial intention of writing a poem.

The second part finds a construction in order to produce this *dénouement*. Here this starting point may be compared to the model figure in geometrical analysis, observable phenomena in an Aristotelian resolution, a physical experimental set-up in the Newtonian analysis, or the scene of crime in a detective story. In all of these cases the desired effect has to be built up from its causes. But in the case of natural and historical phenomena, at least if we are ontological realists about nature and history, the causal ingredients to be found in abduction in a sense already exist in the given totality, and the problem is to find or reveal them, and then the synthesis puts this totality together in an organized way. In the case of poetic creation, the "totality" which serves as the starting point of analysis does not yet contain the ingredients from which it can be composed, but they have to be invented by the artist. Mathematics may be ontologically closer to art than to nature. But the possible differences in ontology do not seem to make a difference in the logical structure of analysis and synthesis within these domains.

Chapter 3
Abduction and Logic

Abstract Since the 1970s Peirce's notion of abduction has become an inspiration for many scholars in formal logic and Artificial Intelligence. This chapter outlines some examples of logical approaches to abductive reasoning. Abduction can be formulated as a rule of inference, which leads to "adaptive logics" of discovery in the family of paraconsistent logics, Abductive Logic Programming, and to non-propositional model-based reasoning (Sect. 3.1). Analysis of the notion of explanation helps to characterize structural rules for abduction (Sect. 3.2). Atocha Aliseda treats the abductive search of missing explanatory premises by means of Beth's semantic tableaux (Sect. 3.3). The GW-model of Dov Gabbay and John Woods argues that abduction is subjunctive and ignorance-preserving (Sect. 3.4). In Jaakko Hintikka's interrogative model of inquiry abduction plays an important role in strategic truth-seeking by questioning (Sect. 3.5).

Peirce's notion of abduction has become an inspiration for many scholars in formal logic and Artificial Intelligence. Without attempting to be exhaustive, we give in this chapter some examples of logical approaches to abductive reasoning: the formulation of abduction as a rule of inference (Sect. 3.1), structural rules for abduction (Sect. 3.2), Aliseda's treatment of abductive search by means of semantic tableaux (Sect. 3.3), the GW-model of Gabbay and Woods (Sect. 3.4), and Hintikka's interrogative model of inquiry (Sect. 3.5).

3.1 Abduction as a Rule of Inference

We have seen in Sect. 1.1 how Peirce introduced the idea of hypothetical or abductive reasoning as an inversion of the Aristotelean Barbara syllogism

(1) $(\forall x)(Fx \rightarrow Gx)$
Fb
$\therefore Gb.$

This deductive inference can be analyzed into two steps:

I. Niiniluoto, *Truth-Seeking by Abduction*, Synthese Library 400,
https://doi.org/10.1007/978-3-319-99157-3_3

$(\forall x)(Fx \rightarrow Gx)$	premise
$Fb \rightarrow Gb$	instantiation
Fb	premise
$\therefore Gb$	Modus Ponens.

Thus, an essential ingredient of (1) is the inference rule known as *Modus Ponens*

(MP) $A \rightarrow B$
A
$\therefore B.$

Another important deductive rule is *Modus Tollens*

(MT) $A \rightarrow B$
$\neg B$
$\therefore \neg A.$

On the other hand, the following inference is not deductively valid:

(AC) $A \rightarrow B$
B
$\therefore A,$

since by the truth–conditions for material implication $\rightarrow$ its premises can true while the conclusion is false. In classical treatments of logic AC is known as the *fallacy of affirming the consequent.*

But if the "secret" of Barbara (1) is *Modus Ponens*,[1] then the logical key of its abductive inversion

(2) $(\forall x)(Fx \rightarrow Gx)$
Gb
$\therefore Fb.$

is precisely the invalid rule AC.[2] Abduction as reverse MP inference thus has its own peculiar logical form (cf. Fann 1970; Walton 2004). Also in the AI circles, abduction has often been interpreted by the rule AC (see Flach and Kakas 2000). However, AC is a "fallacy" only if logic always aims to maximize security in the sense that reasoning is truth-preserving. Indeed, Peirce originally introduced induction and hypothesis as "ampliative" or content-increasing inferences, and continued to emphasize that abduction is needed if we wish to "learn anything" new.

One way of understanding the logic of abduction is to treat it as a form of *plausible reasoning*

---

[1]Peirce himself made this suggestion in 1868 by stating that every deductive syllogism may be put into the form MP, so that hypothetic inference may be called "reasoning from consequent to antecedent" (*W* 2:219–220; *EP* 1:35).

[2]This is of course a simplification, since explanations of the form (1) would cover only what Peirce called "corollarial" reasoning, while more complex examples of explanatory arguments in first-order logic can be "theorematic" (see Sects. 1.3 and 2.3).

(3) A implies B
B
Hence, A is more credible

(see Polya 1957; Walton 2004). We shall explore this idea in a probabilistic framework in Chap. 6. Here it is interesting to note that Peirce's basic model of probabilistic explanation, "simple probable deduction" (1.13), can be understood as a probabilistic version of Modus Ponens:

P(B/A) is high
A
Hence, probably B.

On the other hand, probabilistic Modus Tollens

P(B/A) is high
$\neg$B
Hence, $\neg$A

is not acceptable (see Sober 2008, 49–53).[3]

It is also interesting to see what happens if the abductive rule AC is added as a new rule of inference to classical logic. The answer is that the extended logical framework allows for the derivation of contradictions. In fact, with AC and tautologies of propositional logic we can easily derive from an arbitrary sentence A another arbitrary sentence B:

A	premise
A&B $\rightarrow$ A	tautology
A&B	AC
A&B $\rightarrow$ B	tautology
B	MP

Then by choosing B $= \neg$A, we obtain a proof of the contradiction A & $\neg$A.

Ever since Aristotle, the hallmark of classical logic was the denial of contradictions:

(4) $\neg$(A & $\neg$A).

A logical system violating this principle was thought to be useless, since by the Duns Scotus Law contradiction entails anything (*Ex Falso Quodlibet*):

(5) (A & $\neg$A) $\rightarrow$ B.

However, Fransisco Miro Quesada started in 1976 the study of *paraconsistent logics* which allow contradictions (4) but avoid the undesirable consequence (5) by

[3]Fisher's statistical significance tests were based on probabilistic Modus Tollens, i.e. the null hypothesis H is rejected if the data is very improbable given H, but Neyman and Pearson replaced such tests with comparisons of the likelihoods of rival hypotheses.

restricting applications of some inference rules.[4] Following this idea, Diderick Batens has developed systems of *adaptive logic* which have a dynamic proof theory: reasoning is defeasible and previously derived conclusions may be withdrawn. Meheus and Batens (2006) have studied abduction within an adaptive logic, where the predicative version (2) of the rule AC is added to classical logic, and it is applied as often as possible, but still its applications are restricted by "conditions of abnormality" of the form $(\forall x)(Fx \rightarrow Gx)$ & Ga & $\neg$Fa. In logical derivations it is important to keep track of formulas which depend on the falsity of such abnormality conditions, and rules of inference specify strategies for revoking conclusions whose condition has been defeated (see also Meheus 2004, 2005).

The example of adaptive logic shows that the abductive rule of inference AC can be added to classical logic, so that a paraconsistent logic is obtained. By avoiding contradictions, these systems may generate interesting new conclusions from accepted premises, so that Batens and Joke Meheus propose them as a formalization of a heuristic "logic of discovery" (cf. Chap. 5).

Another line of approach is to replace *Modus ponens* MP by *Disjunctive Syllogism*

(DS) $A \rightarrow C$
$B \rightarrow C$
$A \vee B$
$\therefore C$,

and to study abduction as the inversion of DS:

(6) $A \rightarrow C$
$B \rightarrow C$
C
$\therefore A \vee B$.

Again, (6) is not logically valid, but it applies to situations where we have two alternative conditions for a fact.[5] For example, I see in the morning that grass is wet (C), and this can follow from two assumptions: it was raining last night (A) or the sprinkler was on (B) (cf. Psillos 2000). Then from C we can abductively conclude A or B. If there are good reasons to prefer one of these alternatives, then we can conclude that one as the best explanation of C. This idea of IBE is discussed in Chap. 7.

Douglas Walton (2004) has studied defeasible forms of *Modus Ponens*, where the premise '$A \rightarrow B$' is replaced by 'As a rule, if A then B'. For example, a statement like 'Typically birds fly' allows exceptions (e.g. penguins). Then the inference from 'this is a bird' to 'this can fly' is ampliative, and Walton proposes models of

---

[4]For a survey of paraconsistent logics and their connection to abduction, see Bueno-Soler et al. (2017).

[5]Meheus and Batens (2006) call, in a somewhat peculiar way, adaptive logics with the rule (6) "practical abduction", while "theoretical abduction" allows the derivation of the disjuncts at least as possibilities.

"forward abduction" on this basis as alternatives to (3). However, the connection of this proposal to Peirce's idea that abduction seeks explanations of surprising facts remains unclear.

Other possibilities of reformulating MP include the replacement of the material implication $\rightarrow$ by necessary (strict or nomological) implication or causal condition.[6] The corresponding inversions would then be

(7) Necessarily A implies C
C
$\therefore$ A

and

(8) A causes C
C
$\therefore$A.

The pattern (7) could be studied in modal logic with an operator logical or nomic necessity, while (8) involves causal modalities. Examples of the inverse inference (8), which formalizes Peirce's retroduction, are studied in Chap. 4.

Instead of rules of inference like AC, (6) – (8), the principle of abduction can be formulated by metalinguistic statements. If $\vdash$ is used as a symbol of logical entailment (deduction), then the following expresses an important case of abduction:

(9) A$\vdash$ C
C is true
Hence, there is reason to suppose that A is true.

If A E C means that A explains C, then we have

(10) A E C
C is the case
Hence, there is reason to suspect that A is true.

The patterns (9) and (10) follow the formulations of abduction that Peirce himself gave (see (1.17) and (1.18)). Therefore, it is important to give a more detailed analysis of the relation of explanation E (see the next section).

In the AI treatments, which implement reasoning by computer programs, the premise A$\vdash$ C has been replaced by computational processes with A as input and C as output. For example, in Abductive Logic Programming, a Prolog program checks whether a query C (an instance of some predicate) is derivable from rules, facts, and

[6]For a system of causal abduction, where the arrow $\rightarrow$ is taken from the conditional logic of Chellas, see Beirlaen and Aliseda (2014). This framework avoids abductive conclusions which are self-explanations, contradictory explanations or overinformative explanations.

constraints (relations between predicates). If not, the query is not solvable, and the abductive solution is sought by adding new clauses A, so that C is computable from A.[7]

The idea of abduction has been generalized to non-propositional contexts as well. Paul Thagard and C. Shelley (1997) have replaced sentences with pictures whose visual information is able to produce the data. If perception is a limiting case of abduction for Peirce, this approach suggests a way of formalizing inferences in pattern recognition, machine learning, and model-based reasoning (see Magnani 2001). Interesting applications of this idea have been developed in Magnani et al. (2015) and Park (2017b).

3.2 Structural Rules for Abduction

In her Ph.D. dissertation at Stanford University (supervised by Johan van Benthem), Atocha Aliseda (1997) took up the logical task of analyzing abductive reasoning by means of structural rules. She gives a powerful structural characterization of the "forward" explanatory reasoning from a theory to observational data (see also Aliseda 2003), but one should also ask whether there are any interesting structural rules for the "backward" abductive reasoning from observations to explanatory theories (see Niiniluoto 2007).

Aliseda's (2006) Chap. 3 gives a structural characterization of "consistent explanatory abduction" $\Theta|\alpha \blacktriangleright \varphi$, defined by the conditions (i) $\Theta, \alpha \vdash \varphi$, and (ii) Θ, α are consistent. Here Θ is a theory, α an initial condition (the explanans), and φ the explanadum. In Chap. 5, Aliseda discusses Hempel's DN-model of scientific explanation as a form of consistent explanatory abduction $t|c \blacktriangleright e$, where it is also required that (iii) not $t \vdash e$, and (iv) not $c \vdash e$. *Deductive explanation* differs from mere deduction $\vdash$, since it includes additional conditions which exclude trivial cases of inconsistencies, ad hoc explanation and self-explanation. Hence, the relation $\blacktriangleright$ fails to satisfy conditions of monotonicity and cut, which are typical of deduction $\vdash$, but restricted or modified forms of these principles may be valid.

Aliseda's HD Monotonicity allows consistent strengthening of an explanatory theory:

$$\text{(11)} \quad \frac{t|a \blacktriangleright e \quad t,b|d \blacktriangleright e \quad t,b,a \text{ is consistent}}{t,b|a \blacktriangleright e}$$

The HD Cut principle

[7]For a survey, see Konolige (1996). References to AI studies of abduction are given by Aliseda (2006). See also Romdhane and el Ayeb (2011).

$$(12)\quad \frac{t|a \blacktriangleright b \qquad t|b \blacktriangleright e \qquad \text{not } a \vdash e}{t|a \blacktriangleright e}$$

states that explanation is almost transitive. Other valid principles include Disjunction of Theories

$$(13)\quad \frac{t_1|c \blacktriangleright e \qquad t_2|c \blacktriangleright e}{(t_1 \vee t_2)\,|c \blacktriangleright e}$$

and Conjunction of two Explananda

$$(14)\quad \frac{t|c \blacktriangleright e_1 \qquad t|c \blacktriangleright e_2}{t|c \blacktriangleright (e_1 \,\&\, e_2)}$$

But if explanation is required to be minimal (i.e., without irrelevant additional parts) or preferred (i.e., the best among rival explanations), such structural principles cannot be defended.

To simplify notation, let tEe mean that "t deductively explains e", where the explanans t may be a conjunction of theoretical assumptions and initial conditions. Conditions (iii) and (iv) are special cases of Raimo Tuomela's (1973) non-comparability requirement: in deductive explanation the explanandum should be logically independent of the ultimate truth-functional conjunctive components of the explanans. Let us denote this condition of non-comparability by inc(t,e). Let cons (t,e) mean that t & e is consistent. Then deductive explanation can be defined by

(E) $tEe =_{df} (t \vdash e)$ & cons(t) & inc(t,e).

Tuomela's non-comparabity condition makes explanation irreflexive (cf. Aliseda 2006, 138), so that even Aliseda's Conditional Reflexivity (*op. cit.*, 76) fails:

(15) *Irreflexivity*: not tEt.

Irreflexivity entails that deductive explanation is asymmetric, since eEt and tEe would imply that e and t are logically equivalent:

(16) *Asymmetry*: If tEe, then not eEt.

The principle (11) can now be rewritten in the form:

(11′) *Monotonicity*: If tEe, inc((t & b),e) and cons(t,b), then (t & b)Ee

This implies

(17) *Conjunction of Explanantia*: If t_1Ee, t_2Ee, and cons(t_1, t_2), then (t_1 & t_2)Ee.

Conditions (12)–(14) can now be rewritten:

(12′) *Transitivity*: If uEt and tEe, and inc(u,e), then uEe
(12″) *Right Cut*: If (u&t)Ee and uEt, then uEe
(13′) *Disjunction*: If t_1Ee and t_2Ee, then (t_1 v t_2)Ee

(14′) *Conjunction of Explananda*: If tEe_1 and tEe_2, then $tE(e_1 \& e_2)$.

A similar structural approach has been applied in the study of inductive reasoning, where the relation eIh means that "h is inducible from e" (either in the sense of inductive confirmation or inductive acceptance) (see Hempel 1965; Niiniluoto 1972; Niiniluoto and Tuomela 1973; cf. Sect. 6.1). For example, I is not monotonic and not transitive.

It is interesting to apply the same method to the relation of abduction, understood as reasoning from a fact to one of its potential deductive explanations. The stronger sense, where inference is to the *minimal* explanation or the *best* explanation of a fact, would need a more elaborate treatment. Statistical explanations are also excluded here. Then a simple partial explication of Peirce's schema of abductive reasoning (see (1.18)) would be *converse deduction*:

(CD) $eCDh =_{df} h \vdash e \ \& \ \text{not} \vdash e.$

This definition is known also as "deductive support" (Niiniluoto and Tuomela 1973, 221; Kuipers 2000), and it is a candidate for Smokler's (1968) notion of "abductive inference".

Another partial explication, which is more faithful to Peirce's schema (10), is the definition of abduction as *converse deductive explanation*: h is *abducible* from e iff h deductively explains e. In other words,

(AB) $eAh =_{df} hEe.$

This definition has been proposed in Niiniluoto and Tuomela (1973), 227.[8] The relation A of abducibility trivially satisfies a converse explanation principle:

(18) If hEe, then eAh.

Relation A does not generally satisfy the converse consequence principle

(19) If eAh and $g \vdash h$, then eAg,

since deductive explanation is not always transitive. But by (12′) we have a conditional transitivity property for A:

(19) If eAh and hAu, and inc(u,e), then eAu.

This rule implies a conditional converse consequence principle for A:

(20) If eAh and uEh, and inc(u,e), then eAu.

Abductive inference is non-monotonic, since the following principle is not valid:

(21) If eAh, then (e & b)Ah.

[8]Nepomuceno-Fernández and Soler-Toscano (2007) define a relation of converse explanation A € B by the condition that B entails all elements of A, and give structural rules for €. In contrast to our irreflexivity (15), they accept self-explanations and thus allow reflexivity.

Namely, if h explains e, then it is not always the case that h explains (e & b), even when b is consistent with h. The latter task of explanation is clearly more demanding then the former. But by (14′) the following conjunction principle is valid:

(22) If e_1Ah and e_2Ah, then (e_1 & e_2)Ah.

Another conjunction principle, which is not generally valid for inductive support I, is problematic:

(23) If eAh_1 and eAh_2, then eA(h_1 & h_2),

since the conjunction of two rival explanations may be inconsistent. By (14') we have instead of (23):

(24) If eAh_1, eAh_2, and cons(h_1,h_2), then eA(h_1 & h_2).

Further, principle (13′) implies that

(25) If eAh_1 and eAh_2, then eA(h_1 v h_2).

Most of the principles for A given here would fail, if the notion of abduction would also cover cases of converse inductive explanation.

3.3 Abduction by Semantic Tableaux

Another logical tool which Aliseda (2006) successfully applies in her study of abduction is Beth's semantic tableaux. This tree method was independently discovered by Evert Beth and Jaakko Hintikka in 1955. It provides a powerful search strategy for finding formulas or "abducibles" which serve as the explananda of explanations (cf. Aliseda 2007).[9]

The Beth tableaux give a method of checking whether a sentence C logically follows from a set of premises Σ by trying to find model for the class Σ U {¬C}. Rules for constructing such a model are given in terms of a tableau which is a binary tree with branches depending on the logical connectives in the sentences in Σ U {¬C}. Conjuncts are added to branches, while disjunctions lead to alternative branches. Implications A → B are treated as disjunctions ¬A v B. If the tableau closes (i.e. every branch contains an atomic formula and its negation), the initial set is unsatisfiable or inconsistent, and the deduction Σ⊢ C is valid. If the completed tableau includes open branches, the deduction is not valid. But the structure of the open tree helps to seek an additional premise α such that Σ, α ⊢ C holds, so that the method works as enthymeme retention (cf. Woods 2007).

For example, the validity of Modus Tollens MT is shown by the following tableau with closed branches (indicated by ©)

[9]The first application of the tableaux method to abduction was given by Mayer and Pirri (1993).

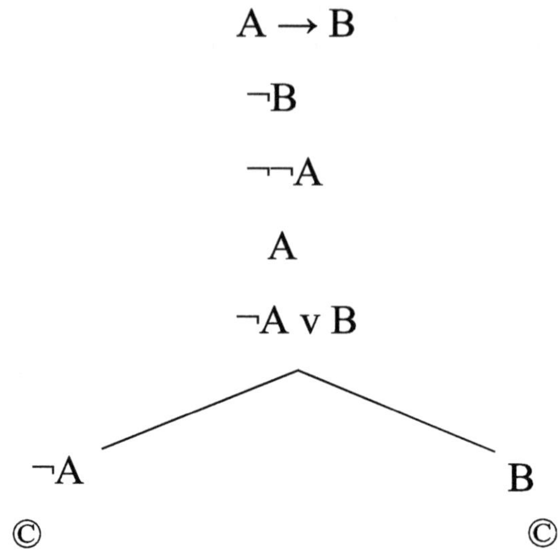

The application to the problem of finding deductive explanations is now straightforward. We start from a background theory Σ and a surprising fact C which is a "lacuna" in the sense that Σ alone does not entail C. We seek to supplement Σ with a new condition A such that the conditions of Aliseda's consistent explanatory abduction $\Sigma | A \blacktriangleright C$ are satisfied, i.e. (i) $\Sigma, A \vdash C$, (ii) Σ, A is consistent, and (iii) not $\Sigma \vdash C$, not $A \vdash C$. This can be illustrated by Disjunctive Syllogism DS: the sentence q is not derivable from the premises $p \rightarrow q$ and $r \rightarrow q$, as the following tableaus shows

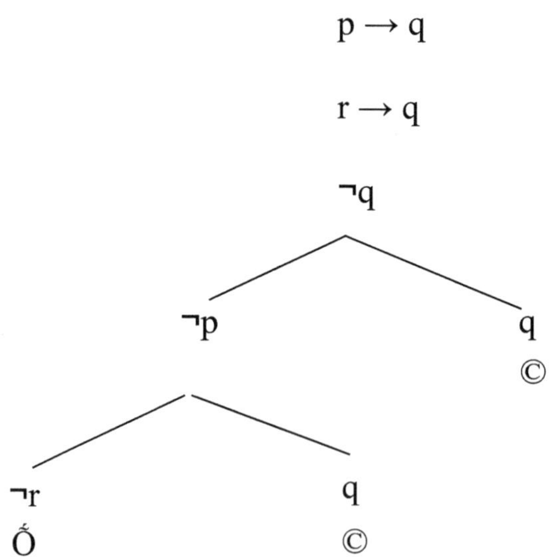

The open branch indicated by Ő can be closed in two ways: by adding p or by adding r. These two abducibles serve as missing premises of explanations, so that by the disjunctive principle (13) their disjunction p v r is also sufficient for the derivation of q (cf. DS).

It is also important to extend the tableaux method to predicate logic, in order to satisfy Hempel's condition that an adequate explanation has to include general laws in its premises. The tableau rule for universal statements adds its instances for all individual constants in the sentences of the branch, and for existential statements its instance for a new individual constant (see Aliseda 2006, 123). Illustration for Peirce's singular abduction (2) can then be given as follows. Suppose we know that $(\forall x)(Fx \rightarrow Gx)$ and we have observed Gb. What is the missing premise? Gb is not derivable from $(\forall x)(Fx \rightarrow Gx)$, since the corresponding tableau has an open branch (indicated by Ő):

$(\forall x)(Fx \rightarrow Gx)$

$\neg Gb$

$Fb \rightarrow Gb$

$\neg Fb$ $\qquad\qquad$ Gb

Ő $\qquad\qquad$ ©

The open branch can be closed by adding Fb as the missing premise, so that the valid deduction (1) is reached. Also stronger sentences with Fb as a conjunct would do the job, but Fb is the minimal abduction which satisfies the desideratum.

Aliseda's (2006) Chap. 8 adds the important observation that a "surprising fact" as the trigger of abduction can be either *abductive novelty* or lacuna (i.e. a statement C not derivable from theory Σ but compatible with Σ) or *abductive anomaly* (i.e. a statement C inconsistent with Σ). In the former case, the sought additional premise A can be added to Σ by expansion. But in the latter case, the incorporation of the new explanation requires that theory Σ is revised. Thus, first theory Σ is revised into Σ' so that Σ' does not any more entail $\neg C$, and then an explanation A is sought so that Σ' with A explains C.[10] In this way Aliseda is able to connect the logic of abduction to the AGM model of belief revision, developed by Peter Gärdenfors (1988). Abductive belief revision is discussed in Sect. 8.5 below.

[10]Kuipers (1999) adds a variant where the revised theory Σ' is obtained from Σ by the process of concretization. Cf. Chapter 8.

Aliseda's notions of abductive problem and abductive solution can be given precise formulations within the framework of dynamic epistemic logic (see Nepomuceno-Fernández et al. 2017). Using plausibility models, which include a preference ranking of possible worlds, this approach can also identify and select the explanation which is the best relative to the agent's epistemic state.

Aliseda's logical approach can be supplemented by elaborating the notion of an abductive problem within a probabilistic framework. A fact C is surprising if its probability $P(C/\Sigma)$ given our background theory Σ is low.[11] Then expanding Σ with an additional premise A should increase the probability of C so that $P(C/\Sigma\&A)$ is high (or at least higher than $P(C/\Sigma)$). In this case abduction leads to an inductive-probabilistic explanation, which in the limit with $P(C/\Sigma\&A) = 1$ reduces to a deductive explanation. In this way one can give plausible explanations of some accidental or unexpected events by showing how they were possible (cf. Hempel 1965, 428). In the case of abductive anomaly, where C seems to be impossible given our theory Σ, C is shown to be possible if its probability $P(C/\Sigma'\&A)$ given the revised theory Σ' and the new explanation A is greater than zero. Again, in the limit $P(C/\Sigma'\&A) = 1$. Such *how possible* –explanations, which are weaker than *why necessary* –explanations, may refer to historical courses of events or causal mechanisms which serve to show how certain factual situations have emerged ('How come this city center is so ugly?') or how some individuals ('How can Usain Bolt run 100 meters in less than 10 seconds?') or species ('How are the cranes able to migrate in the winter to the south?') have their surprising capacities.[12]

3.4 The GW Model of Abduction

The work of Dov Gabbay and John Woods (2005) has recently received a lot of attention in the theory of abduction (see e.g. Magnani and Bertolotti 2017). But their model has some peculiar features which distinguish it from most other contributions in the Peircean tradition. Woods (2007) calls schemas (9) and (10), or their refinements, the AKM model of abduction, where the letters may refer to Aliseda, Kowalski, Kuipers, Kakas, and Magnani, while their own alternative is called the *GW model*.

The GW model starts from an "ignorance problem" with respect to some proposition T, where the available knowledge base in insufficient to "attain" T or to solve whether T is the case or not. This is clearly different from Peirce's account of abduction, which starts from a situation where a surprising fact C is known, and we seek an explanation for C. Here we are ignorant about *why* C is the case, but not *whether* C is the case. In other words, abductive inference is triggered by a surprising

[11]In a fair lottery, the probability that a given single ticket wins is low, even though it is not surprising that at least one among many voters wins.

[12]For how possible explanations in the theory of evolution, see Sect. 4.4.

fact C (a novelty or an anomaly), so that we already know that C is the case. Our cognitive target is not to look at additional reasons for knowing *that* C, but we wish to know to know *why* C is the case. Thus, the GW model fails to make a distinction between explanation-seeking and reason-seeking why-questions (see Hempel 1965, 334–335). It proposes to extend the scope of abduction from explanatory situations to all other epistemic problems. But it would be misleading to include e.g. predictive problems as instances of abduction. Therefore I think we should agree with Hoffmann (2010) and Schurz (2017) that the idea of non-explanatory abduction overstretches the notion of abductive reasoning.

Another important feature of the GW model is its thesis that abduction is "ignorance-preserving". An abductive hypothesis H attains the cognitive target T only subjunctively or presumptively, and in "partial abduction" H can be presented as a conjecture and in "full abduction" it will be activated in action. Thus, in abduction the agent's "ignorance remains but he is not overcome by it" (Woods 2017, 139). Woods (2017) makes it clear that his intention is to develop a sort of non-evidentialist epistemology, where "knowledge" can be attained in the absence of evidence. When a hypothesis H is introduced to explain C, we are initially ignorant about the truth value of H, but in my view the question whether the ability of H to explain C gives some support to H should be investigated separately (see Chap. 6 below), and not be taken as a stipulation about "ignorance-preserving". We shall comment on this view of knowledge and rational action in Sect. 5.3, and in Chap. 6 we argue that successful explanation of facts C by a theory A increases the epistemic probability of A. So in our view abduction is credence-increasing, and we have reason to disagree with e.g. Magnani who thinks that "the GW-model does a good job in modeling the ignorance-preserving character of abduction" (see Magnani et al. 2015, 132).

Appealing to Peirce's formulation of the premise of (1.18), i.e. "If A were true, C would be a matter of course", Woods (2007) argues that the AKM model lacks a subjunctive consequence relation. The GW model, instead, assumes only that *were* A the case then the desired cognitive target would be attained, and therefore non-subjunctive DN-explanationism is not really a case of abduction. But in schemas (9) and (10) A is not indicatively asserted as a fact but as a hypothesis. The force of the truth-preserving deductive consequence relation $A \vdash C$ can be stated by a subjunctive: if A were true, then C would be true, and Peirce surely knew this when he chose his words. Moreover, Peirce continued to link abduction to explanation, and it is clear that in the schema (10), i.e.

A explains C
C is the case
Hence, there is reason to suspect that A is true,

the notion of explanation has to be potential explanation (with unknown truth value). If A E C would presuppose that the explanans A is true, then in (10) the inference to the conclusion would be logically valid (even without the second premise). Thus, the real difference between the AKM and GW models is not in the question whether consequence is subjunctive or not, but rather in the further controversial GW claim that abduction is "ignorancc-preserving".

3.5 Hintikka's Interrogative Model of Inquiry

After seminal studies on inductive logic in the 1960s,[13] Jaakko Hintikka extended his 1962 epistemic logic to a logic of questions in the 1970s. In the next two decades he developed an interrogative model of inquiry, which stresses the importance of questioning in rational knowledge-seeking (see Hintikka 1999). In his "Socratic epistemology", all of our knowledge consists of answers to questions (see Hintikka 2007). In this logic of discovery, induction does not play any definite role besides deduction, but abduction has an important function as an "interrogative move".

Hintikka's starting point is similar to Aliseda's (2006) work on abduction: we are given a body of background knowledge Σ from earlier inquiry, and this previous information can be used as a resource to solve some cognitive problem. If the problem is whether C is true or not, we try to derive C or ¬C from Σ. This inquiry includes two kinds of steps or moves:

(i) logical inference steps
(ii) interrogative steps.

Here (i) means reasoning by deductive logic from previous knowledge, so that in Hintikka's model of inquiry all inferences are deductive. If the desired conclusion cannot be derived from Σ, we again face the problem of finding a missing premise. But instead of using a technique like semantic tableaux to calculate the missing premise, with interrogative steps we bring in new information by questioning. This is the Socratic element in critical inquiry. Questions are requests information, and they are answered by "oracles", which may include observation by naked senses or instruments (telescopes, microscopes,. . .), experiments as questions put to nature, witness testimony in a court of law, patient at a diagnostic interview, memory and tacit knowledge, data bases in computers (data mining), and internet (searching by Google) (see Hintikka and Bachman 1991). But in the context of scientific investigation, the main oracle is the actual world or some part of it, so that inquiry can be described as a game between the *Inquirer* and *Nature*, and the most important questions are experiments (Hintikka 1999, 161).[14] While in the general case all information given by such oracles is not correct, one needs a critical stance and special caution in evaluating the answers. But in his basic model of science, Hintikka assumes that Nature's answers are true and known to the Inquirer.

---

[13]For a survey, see Niiniluoto (2011c).

[14]The idea that experiments are questions to nature goes back to Francis Bacon and Kant (see Sintonen 2009).

A dynamic aspect of inquiry arises from the principle that the Inquirer may ask only those questions whose *presupposition* is known. For example, the presupposition of the explanatory why-question "Why p?" is that p is true. The presupposition of the propositional question "Which one of the propositions $p_1, ..., p_k$ is correct?" is the truth of $p_1 \vee \ldots p_k$. In a yes-or-no-question these alternatives are p and $\neg$p. The presupposition of the WH-question "Who lives here?" is that someone lives here, and the answer should guarantee that the questioner knows who lives here. In this way, the logic of questions is embedded in epistemic logic.

Hintikka distinguishes definitory and strategic rules: In a game like chess, there are *definitory rules*, which describe how to play the game correctly (what moves are allowed for different pieces), and *strategic rules*, which give insight into how to play the game successfully and well (what are effective methods of defending your position, attacking the opponent, and winning the game with a mate).[15] Similarly, in inquiry as a game for finding new knowledge, definitory rules are described by the notion of logical consequence, while strategic rules are given by methods of knowledge-seeking by questioning.

Hintikka gives special emphasis to the art of questioning, with ample illustrations from Arthur Conan Doyle's Sherlock Holmes stories (see Hintikka and Hintikka 1983). His favourite example is the curious incident of the racing horse Silver Blaze who has been stolen at night from its stable. There was a watchdog in the stables. Holmes notes a curious incident of the dog by raising the question: did it bark during the night? The only person any watchdog does not bark at is its master. Therefore, Holmes concludes, the thief was the stable master.

The interplay of such rules can be illustrated by the following example. You try to show that C is true, but you cannot logically derive it from your background information K. Analysis of the inference shows that the missing premise A would allow the derivation of C from K&A. You ask from your oracle whether A is the case. If the answer is yes, you have an argument for C.

Hintikka's interrogative model, which tries to find the most effective strategic ways of knowledge seeking by series of deductive inferences and questions, has no special role for ampliative inferences like induction and abduction. In particular, abduction as inference to the best explanation is not needed, but rather "abductive 'inferences' must be construed as answers to the inquirer's explicit or (usually) tacit question put to some definite source of answers (information)" (Hintikka 2007, 49). Hintikka relates this notion of abduction to Peirce's statements about the role of abduction as interrogation (*CP* 6.526) and as the only logical operation which introduces new ideas (*EP* 2:216). But he also argues that Peirce failed fully to appreciate the fact that the success of abduction should be assessed from the strategic perspective.

[15]The reviewer of this book points out that the distinction between definitory and strategic rules corresponds to the fourteenth century philosopher William Burley's distinction between rules that "constitute the practice" and the rules that "pertain to its being practiced well".

Hintikka's proposal to treat abduction as an interrogative step in inquiry is in many ways attractive. But if one extends the notion abduction by relating it to all kinds of questions, including WH-questions, we should keep in mind that for Peirce abductive reasoning starts from the problem of explaining some surprising fact. Hintikka's approach can be defended by noting that an inquiry aiming at explanation usually needs to investigate many facts which serve as either initial conditions or explanantia of explanatory arguments. As illustrated by problems of crime, such additional information may be sought by what-, where-, when-, who-, how-, and why-questions (cf. Sect. 2.4). Still, it would be misleading to state that all WH-questions, which function as parts of a broader process of inquiry, are abductive. Further, if abduction is restricted to explanatory why-questions, then the metaphor of questioning Nature – which is already problematic for propositional questions with theoretical alternatives – starts to break down: explanation-seeking why-questions are not reducible to yes-or-no-questions, since they do not offer Nature a ready-made list of alternatives to choose from (see Sintonen 2009, 207).

Hintikka's interrogative model can be developed in several directions.[16] One is to allow that the information source can give uncertain answers, so that observational statements are correct only with some probability (see Niiniluoto 1997). In his later work Hintikka became again interested in the problem of induction, as he conceded the point that in typical curve-fitting problems the function relating the measured variables is not given by Nature but has to be discovered by the Inquirer (see Niiniluoto 1997, 122). But if inductive generalization is permitted as a logical inference step along deduction, then the same should go with abduction. Extending the interrogative model with these ampliative inferences – with lots of examples in science and everyday reasoning – would not diminish its power to explicate the important dynamic aspects of human reasoning.

Hintikka's interrogative model can be compared to the treatment of argumentation and abduction in dialogical logic, which is inspired by his game-theoretical analysis truth in terms of winning strategies in a dialogical game. A good presentation is given by Barés Goméz and Fontaine (2017), but in my view their article takes too seriously the GW model and its ignorance thesis (see Sect. 3.4). This is seen in the fact that they try to characterize the proper speech act for abductive solutions which are "conjectured as being plausible without any commitment to the belief of the truth of what is expressed". But they don't find any alternative between mere questions and assertions, so that the problem is left unsolved. This kind of impasse is avoided in dynamic epistemic logic, which distinguishes knowledge and belief, and the agent has the right to *believe* the abductive solution (see Nepomuceno-Fernández et al. 2017, 274).

[16]Hamami (2015) extends Hintikka's model to a multi-agent setting, which has been studied in dynamic epistemic logic.

Chapter 4
Inverse Problems

Abstract In this chapter we give illustrations of abductive reasoning that Peirce called "retroduction". His own paradigm example, the inference to the existence of Napoleon Bonaparte from present documents and monuments, can be generalized to a method of evaluating historical hypotheses by means of the causal traces of past events (Sect. 4.1). Peirce's thesis about perception as an extreme case of retroduction needs special attention. We argue that the new growing branch of applied mathematics called inverse problems deals successfully with various kinds of abductive inference within a variety of scientific disciplines (Sect. 4.2). The fundamental theorem about the inverse reconstruction of plane functions from their line integrals was proved by Johann Radon already in 1917. The practical applications of Radon's theorem and its generalizations include computerized tomography which became a routine imaging technique of diagnostic medicine in the 1970s (Sect. 4.3). It is further argued that in biology the reconstruction of the evolutionary tree of life on the basis of present evidence is a retroductive task in Peirce's sense (Sect. 4.4). Starting from the 1860s, the idea of evolution was applied in the study of culture (philology, ethnology, anthropology, folkloristics), with the aim of reconstructing family trees of languages and stemmas of oral and written texts (Sect. 4.5). The inferential abductive structure of textual criticism and stemmatology in cultural sciences is shown to be similar to cladistics in biological phylogenetics.

In this chapter we give illustrations of reasoning that Peirce called "retroduction". His own paradigm example, the inference to the existence of Napoleon Bonaparte, can be generalized to a method of evaluating historical hypotheses, but perception as an extreme case of retroduction needs special attention (Sect. 4.1). We argue that the new growing branch of applied mathematics called inverse problems deals successfully with various kinds of abductive inference within a variety of scientific disciplines (Sect. 4.2). The fundamental theorem about the inverse reconstruction of plane functions from their line integrals was proved by Johann Radon already in 1917. The practical applications of Radon's theorem and its generalizations include computerized tomography which became a routine imaging technique of diagnostic medicine in the 1970s (Sect. 4.3). It is further argued that in biology the reconstruction of the evolutionary tree of life on the basis of present evidence is a retroductive

I. Niiniluoto, *Truth-Seeking by Abduction*, Synthese Library 400,
https://doi.org/10.1007/978-3-319-99157-3_4

task in Peirce's sense (Sect. 4.4). Starting from the 1860s, the idea of evolution was applied in the study of culture (philology, ethnology, anthropology, folkloristics), with the aim of reconstructing family trees of languages and stemmas of oral and written texts (Sect. 4.5). The inferential abductive structure of textual criticism and stemmatology in cultural sciences is shown to be similar to cladistics in biological phylogenetics.[1]

4.1 On Retroduction

For Peirce, abduction is inference which proceeds from effects to their causes (singular abduction) or from surprising observations to their tentative explanation (theoretical abduction) (see Chap. 1). *Retroduction* is a variant of singular abduction, where the general premise is a causal law of succession: if we know or assume that all Fs are followed by G, then from the event Ga we can reason backward in time to the conclusion Fa. Detective stories, which try to identify a murderer on the basis of clues from present evidence, use this kind of backward reasoning.[2]

Peirce himself illustrated retroduction by two examples in 1878:

> Fossils are found; say, remains like those of fishes, but far in the interior of the country. To explain the phenomenon, we suppose the sea once washed over this land.
>
> Numberless documents and monuments refer to a conqueror called Napoleon Bonaparte. Though we have not seen the man, yet we cannot explain what we have seen, namely, all these documents and monuments, without supposing that he really existed. (*CP* 2.625; *EP* 2:54)

The inference from present testimony of memoirs, monuments, and public documents to the existence of Napoleon Bonaparte (and to a fuller narrative of his life) suggests that all historical sciences apply abductive methods: past events leave traces as their causal effects, and from these traces we make retroductive inferences to the past. Indeed, our only access to the past is through abductive hypotheses which explain the present evidence.[3]

---

[1]Sections 4.2 and 4.3 of this chapter are mainly based on Niiniluoto (2011a). Earlier versions of Sects. 4.4 and 4.5 were first presented in a symposium "Mind and Evolutionary Inference" at the University of Helsinki on August 18, 2014, and in EPSA15 in Düsseldorf on September 25, 2015.

[2]Edgar Allan Poe's detective stories were composed so that they have only one backward solution (see Sect. 2.4).

[3]This can be used as an argument for the reality of the past. To use Bertrand Russell's example, even though we cannot any more know whether Winston Churchill sneezed a hundred times in 1939, this statement has an unknown truth value, based on facts about his life. In other kinds of cases, sufficient information about past events has been preserved in causal processes, and the claims about them can be known abductively. See Niiniluoto (2014a). Turner (2007) gives a survey of anti-realism about the past.

Similar examples of everyday abduction can be easily multiplied. I am looking a photo of my father as a young adult among his family members. Taking clues from their outlook and clothes, I make an initial estimate of the decade when the photo has been taken. From the fact that an elderly lady is sitting in the middle with a bunch of flowers, I make the reasonable guess that the family is celebrating my grandmother's 60th birthday.

Carol Cleland (2011) argues – without reference to Peirce's retroduction – that historical hypotheses are assessed primarily by their capacity to give common cause explanations about puzzling associations between traces. While Elliott Sober (1988) thinks that by default common cause explanations and separate cause explanation should be treated neutrally as equally plausible, Cleland states that there is strong empirical evidence for the prevalence of common causes, since typically past events are overdetermined by their effects. For example, a volcano leaves numerous and diverse effects, and we need not know all of them to identify the cause.[4] Besides human historiography (including archeology and anthropology), such abductive methods are used in (pre)historical natural sciences like cosmology, astronomy, geology, and paleontology. In the next sections of this chapter we shall see that similar examples abound in medicine, biology and cultural sciences.[5]

Peirce also mentions perceptual judgment as an extreme, "not controllable" and "not fully conscious" case of abduction (*CP* 5.181). It is a "judgment absolutely forced upon my acceptance", but still as the limiting case of abduction it is based on hypothetic inference, which can be formulated as an inverted syllogism (*CP* 8.64).[6] This can be understood in terms of Peirce's realism: the fundamental hypothesis of science is that "there are Real things, whose characters are entirely independent of our opinions about them", and

> those Reals affect our senses according to regular laws, and, though our sensations are as different as are our relations to the objects, yet, by taking advantage of the laws of perception, we can ascertain by reasoning how things really and truly are. (*CP* 5.384)

Typically surprising perceptions arise, when we habitually act upon our beliefs and doubt leads to new inquiry (*CP* 5.12). Thus, we observe real things in the external world, and our "percepts" (involving their qualities, resistance against our will, and

---

[4]Adrian Currie (2017) points out that historical sciences, besides analyzing present traces, also exploit dependencies between past events themselves, but this does not change the fact that ultimately knowledge about the past has to rely on abduction.

[5]Mill (1906) also suggests that "the historical method" uses the "inverse deductive method", but he does not develop it in the same way as Peirce. For Whewell's philosophy of historical science, see Quinn (2016a).

[6]In their AI approach, Josephson and Josephson (1994) analyze perception as an example of abductive reasoning. Its special case is understanding spoken language, where the intended meaning of an utterance is interpreted by abduction (see also Hobbs 2004). More generally, Eco (1990) argues that Peirce's abduction is "the logic of interpretation". The hermeneutic reconstruction of meaning should be distinguished from the application of abduction to the assessment of the quality of testimony: in Lipton's (2007) Testimonial IBE, the truth of an utterance is the best explanation of its utterance by a witness.

generality) lead to perceptual judgments (see *EP* 2:62). In this way, perception involves (without our recognizing this) hypothetical inference from effects to causes.[7]

In such special cases, abduction is "perfectly certain" or "compelling". In spite of his fallibilism, Peirce grants that to entertain a doubt about Napoleon's existence "would surely be downright insanity" (*EP* 2:54).[8] And we cannot conceive any way of denying the perceptual judgment (*EP* 2:229–230), even though perceptual judgments are fallible statements about reality.

It is important to complement Peirce's "fundamental hypothesis" by noting that the causal influence from "real things" to our sensations is not unidirectional, since in experimentation we are in causal interaction with external objects.[9] But this raises the objection that in historical sciences we cannot causally interact with past objects. Derek Turner (2007) suggests that it is more difficult to know unobservables in the past than unobservables in the tiny, since we can manipulate electrons but not dinosaurs. This difference is important, but it may be to some extent compensated by analogical inference based on computer simulations of past circumstances. In any case, arguments about the past are based on retroductive reasoning.

Discussion on retroduction restarted when Carl G. Hempel analysed in 1958 the logical structure of "postdiction" or "retrodiction" (see Hempel 1965, 173). For the former term, he referred to Reichenbach in 1944, for the latter term, to G. Ryle in 1949 and J. M. Robertson in 1895, but Peirce was not mentioned all. For Hempel, postdiction has the same structure as DN-explanation and prediction, viz. derivation of a statement about an event from general laws and initial conditions, with the only difference that in postdiction the event must occur before the argument. Thus, Hempel's postdictions are deductive and not abductive (inversions of deductions) in Peirce's sense.

An elegant analysis of explanation, prediction, and retrodiction in discrete state systems was given by Nicholas Rescher (1970). His inspiration came from Hempel, and again no reference to Peirce's abduction or retroduction was given.[10] Rescher's DS-systems allow both deterministic and probabilistic retrodictions (see the next section).

---

[7]To say, in semiotic terms, that the perceptual judgment is an index of the percept is a form of the causal theory of perception (see Tiercelin, 2005). For Peirce's theory of perception and its emphasis on "the outward clash", see Hookway (1985), Ch. V. Peirce's remarks on perception as a form of abduction can be extended to memory, which causally preserves our perceptual experiences in our mind over time. As a faculty giving access to past events memory is even more fallible than perception.

[8]Yet, such historical existence statements may be mistaken. Tuomas Heikkilä has argued that the old legend about Bishop Henrik who supposedly brought Christianity to Finland in the twelfth century and got killed by an angry peasant, is a mythological figure.

[9]This idea on interaction is in harmony with the pragmatist emphasis on human action, and was developed especially by John Dewey in his criticism of "the spectator theory of knowledge". Manipulation or intervening has been defended by Ian Hacking (1983) as a test of the reality of theoretical entities. See also Niiniluoto (1999a), 40, 94.

[10]Later Rescher has extensively discussed Peirce's work. See, for example, Rescher (1978). Note that in this later work Rescher uses the term "retroduction" for the process of hypothesis testing and elimination.

4.2 Inverse Problems in Systems Theory

When Peirce wrote about abduction, inverse methods had not yet been systematically studied in mathematics. His manuscripts about geometry contain only a brief remark about the "reversion and perversion" of serial orders (Peirce 1976, vol. 2, 316).

Since the 1970s "inverse problems" have been established as a new subdiscipline of mathematics, with its own academic journals (*Inverse Problems* since 1985, *Journal of Inverse and Ill-posed Problems*, *Inverse Problems in Science and Engineering*), international congresses, scientific societies, and centres of excellence. Mathematicians working in this area have not paid attention to Peirce's concept of abduction, however. Likewise, in spite of abundant applications in various domains, the studies of abduction in logic, AI, and philosophy of science do not contain references to this new mathematical research programme. The first discussion on this theme is given in Niiniluoto (2011a).

In this section, I shall give some examples of inverse problems in connection with statistics and systems theory. In the next section, the fundamental theorem proved by Radon in 1917 is outlined and some of its important applications are described.

Simple algebraic treatments of inverse reasoning have been well known in modern mathematics. One method concerns the identification or estimation of parameters on the basis of equations. Let us assume a simple linear model between two physical variables x and y:

(1) $y = ax + b$.

Then parameters a and b can be calculated, if two pairs of values of x and y are known. Geometrically, this means that two points on a plane determine a straight line.

If (1) is a hypothetical theory, then *linear regression analysis* recommends that a and b are chosen so that they minimize the Least Square Distance of the line $y = ax + b$ from the observed data points $<x_1,y_1>, \ldots, <x_n,y_n>$:

(2) $\sum_{i=1}^{n} (y_i - ax_i - b)^2$

The basic idea of this method of least squares, and its connections to observational errors, was developed by Legendre, Gauss, and Laplace already in the first years of the nineteenth century. The statistical method of regression analysis was systematically developed around 1900 by U. Yule, Karl Pearson, and A. A. Markov.[11]

[11]Schurz (2008a, 2017) shows that statistical factor analysis, a generalization of regression analysis, can be understood as a form of abduction.

More generally, assume that variable y depends on other variables $\underline{x}$ through the function $f(\underline{x},\underline{a})$, where $\underline{a} = (a_1, \ldots, a_m)$ are real-valued parameters. Here $f(\underline{x},\underline{a})$ can be a linear or non-linear function. Then the equation

(3) $$y = f\left(\underline{x, a}\right)$$

solves the "normal" or *forward problem*: given values of the independent variables $\underline{x}$ and parameters $\underline{a}$, we calculate or predict the value of the dependent variable y. The *estimation problem* is to use information about $\underline{x}$ and y to identify the parameters $\underline{a}$. This can be done by assuming the form of (3), and by choosing $\underline{a}$ so that the distance of the curve (3) from the observed data is minimal. This inference is sometimes motivated by the idea of maximising predictive accuracy (Forster and Sober 1994). But, as the estimation task identifies the best approximate explanation of the data, it can be considered as a special case of Peircean abduction. The abductive inference rule in this case is the following:

(4) Given observational data E about variables $\underline{x}$ and y, conclude that among hypotheses of form (3) the one giving the best approximate explanation of data E is most truthlike.

(See Niiniluoto 2004; cf. Sect. 8.2)

Another kind of *inverse problem* is to assume knowledge about (3) and $\underline{a}$, and infer the values of $\underline{x}$ from an observed value of y. In other words, in an inverse problem we try to *infer input* $\underline{x}$ *from the model* (3) *and output* y. In the simple linear case, x can be solved from equation (1), where $a \neq 0$, so that $x = (y-b)/a$. This type of inference is clearly a mathematical variant of the Peircean original model of hypothesis or its retroductive version.

A statistical version of the equation (3) adds a random variable ε with an assumed probabilistic error distribution P:

(5) $$y = f\left(\underline{x, a}\right) + \varepsilon.$$

Then (5) allows us to make probabilistic "forward" predictions $P(y/\underline{x, a})$ about y on the basis of $\underline{x}$ and $\underline{a}$. Relative to (5), inverse retroductions give probabilities $P(\underline{x}\,/y, \underline{a})$ of the values of $\underline{x}$, given values of y and $\underline{a}$. Again, statistical methods may be used for estimating the parameters $\underline{a}$ in (5).

In astronomy, a typical example of the forward problem is the use of Newton's laws together with the initial state of the planetary system to calculate future occurrences of eclipses. A famous case of inverse inference is the independent discovery the planet Neptune by J. C. Adams and U. J. J. Le Verrier in 1846. Irregularities in the movement of the planet Uranus were observed, and – instead of rejecting Newton's well-established theory – they postulated the existence of a perturbing body and by Newton's laws calculated its location. The conclusion of this abductive inference was then verified by telescopic search.

Another contemporary example comes from cognitive neuroscience. Russell Poldrack's lab at Stanford University studies forward inferences from a psychological process (e.g. short-term memory) to a brain image, and "reverse inferences" from the activation of a brain region to a cognitive process (see Poldrack 2006). He does not use the term 'abduction', but notes that the reverse inferences are non-deductive.

A special kind of inverse problem concerns the behaviour of closed physical systems. A *discrete state system* S is at time t in one of the states $s_1, ..., s_k$, and lawlike transitions between states satisfy the Markov condition: state at t + 1 depends only on the previous state at t. The system S is *futuristically deterministic*, if for each t the state at t determines the state at t + 1, and *historically deterministic,* if for each t the state at t determines the state at t−1 (cf. Montague 1974). In such cases, "D-predictions" and "D-retroductions", respectively, are possible. A probabilistic state system is characterized by transition probabilities P(s′/s) between successive states. Then the inverse problem about system S concerns the possibility of retroductive inferences. Inverse probabilities of causes, given their effects, can be calculated from transition probabilities by Bayes's Theorem. It can be seen that D-retroduction may be impossible in a futuristically deterministic system, i.e. futuristic determinism does not imply historical determinism. But D-retroduction may be possible in an indeterministic system (see Rescher 1970). A sufficient condition for the equivalence of futuristic and historical determinism is the time-reversibility of the transition laws concerning the system (Earman 1986).

For non-discrete state systems, the states constitute a continuous subset – the state space or phase space – of some normed linear space, e.g. the n-dimensional Euclidean space $\mathbb{R}^n$ ($n \geq 1$) or the infinite-dimensional Hilbert space. Such systems with "ergodic" properties have been studied in statistical mechanics since the late nineteenth century (see von Plato 1994). The condition of "stability" states that the outcome of the system depends continuously on its initial state. In a system which is unstable at state s, in arbitrarily small neighbourhoods of s there are other states s′ such that s and s′ lead to diverging outcomes. It was an important observation that deterministic physical systems can be unstable, so that their behaviour cannot be predicted by finite information about their initial state. Such systems are known as "chaotic" (Earman 1986).

Open dynamical systems are in interaction with their environment via an external input. The independent variables can be divided into external input variables $\underline{x}_I$, which often can be manipulated by us, and internal state variables $\underline{x}_S$. In addition to the output function (3), which now depends on both $\underline{x}_I$ and $\underline{x}_S$, a futuristically deterministic system is characterized by a state function $g(\underline{x}_I,\underline{x}_S)$, which gives the next state of the system with input $\underline{x}_I$ and state $\underline{x}_S$. In probabilistic state systems, the function g is replaced by a physical probability distribution over the states.

Many inverse problems are *ill-posed* in the sense defined by J. Hadamard: the existence and uniqueness of their solutions are not guaranteed. Especially in non-linear systems inverse problems may be unstable in the sense that slight variations in the value of y lead to great differences in the calculated values of $\underline{x}$. In other words, even a historically deterministic system can be chaotic relative to the past. In ill-conditioned inverse problems, the solution $\underline{x}$ could be calculated only from an infinite number of observations. Sophisticated probabilistic, statistical, and computational methods have been developed in order to obtain practical and reliable approximate solutions to such problems (see, e.g. Tarantola 2005).

4.3 Radon's Theorem and Computerized Tomography

Johann Radon (1887–1956), professor of mathematics at the University of Vienna, proved in 1917 an abstract theorem in function theory. According to Radon's theorem, a sufficiently regular plane function can be uniquely reconstructed from its line integrals. This theorem, with its generalizations, has turned out to be a key to the advanced mathematical study of inverse problems and a source of important practical applications in various fields of science and technology.

Radon's idea can be illustrated by starting from the discrete case.[12] Consider a square with a natural number in each of its cells (see Fig. 4.1). Then the forward problem is to count the sums of all rows and columns. The inverse problem starts from the marginal sums and attempts to reconstruct the numbers in the square. So, for example, in a 3 × 3 square there are nine unknown numbers, and from the marginal sums we can write six equations with them. In many cases, such an inverse problem will have more than one possible solution. For example, the same marginal sums are produced by another assignment of numbers (see Fig. 4.2). But if we add three diagonal sums to the square, then the number of variables and equations are the same, and a unique solution is obtained by solving the system of nine equations.

The same inverse problem can be immediately generalized to any n × n square of integers. If n = 100 000, then the number of unknown variables is 10 billion. If

Fig. 4.1 3 × 3 square of integers with marginal sums

1	2	3	*6*
2	3	4	*9*
3	4	5	*12*
6	*9*	*12*	

Fig. 4.2 Alternative 3 × 3 square

6	0	0	*6*
0	9	0	*9*
0	0	12	*12*
6	*9*	*12*	

[12]This way of introducing inverse problems has been used by Lasse Päivärinta, who has been director of the Finnish centre of excellence for inverse problems at the University of Helsinki and the President of the Finnish Inverse Problems Society.

10 billion lines are drawn to intersect the square, so that all the cells of the square are covered, and the sums of integers on each line are calculated, then we have again a system of equations with an equal number of unknown variables. The solution (if any) to such a system can be computed by using the inverse matrix method.

Radon's approach is a further generalization of this problem.[13] Replace integers by real numbers and the finite square by a real-valued function f: $\mathbb{R}^2 \rightarrow \mathbb{R}$ on the infinite two-dimensional plane $\mathbb{R}^2$. Let L be a line in $\mathbb{R}^2$. Then the "sum" of all values f(x,y) of function f on points z = (x,y) belonging to line L is mathematically expressed by the line integral of f over L:

(6) $R(f, L) = \int_L f(z)dz$

Line integrals are generalizations of the ordinary method of integrating one-place real functions with respect to the x-axis. For example, if line L is geometrically represented by the equation $x\cos\theta + y\sin\theta = s$, where s is the shortest distance from L to the origin and θ is the angle L makes with the y-axis, then R(f,L) is

$$\int_{-\infty}^{\infty}\int_{-\infty}^{\infty} f(x,y)\delta(x\cos\theta + y\sin\theta - s)dxdy$$

where δ is Dirac's delta-function. Another expression for R(f,L) is

$$G(s,\theta) = \int_{-\infty}^{\infty} f(s\cos\theta - p\sin\theta, s\sin\theta + p\cos\theta)dp.$$

The function R(f) which gives the integral R(f,L) of function f over all lines L on $\mathbb{R}^2$ is called the *Radon transform* of f.

The inverse problem concerning function f now consists of the question whether f can be uniquely computed from its Radon transform, i.e. from knowledge about its line integrals. For this purpose, Radon assumed that f is sufficiently regular on $\mathbb{R}^2$, i.e. a "smooth", infinitely differentiable and rapidly decreasing function. Let $F_z(q)$ be the mean value of all integrals R(f,L) over lines L passing at the distance q from point z in $\mathbb{R}^2$:

$$F_z(q) = \frac{1}{2\pi}\int_0^{2\pi} G(x\cos\theta + y\sin\theta + q, \theta)d\theta.$$

Then f is uniquely determined by the line integrals through the formula

(7) $$f(z) = -\frac{1}{\pi}\int_0^{\infty} dF_z(q)/q$$

where $dF_z(q)$ is the first derivative of $F_z(q)$ with respect to q.

[13]Radon's original 1917 paper, with historical information, is published in Radon (1987). Derivations of Radon's main result, with its generalizations, are given in the monograph of Natterer (2001).

Radon's inversion formula (7) was generalized to the n-dimensional space $\mathbb{R}^n$ for $n \geq 2$, by Ph. Mader in 1927. In 1955, F. John coined the term "Radon transform" for the function which associates a function f on $\mathbb{R}^n$ with its integrals over hyperplanes, i.e. linear subspaces of $\mathbb{R}^n$ with dimension n–1.[14] The function which associates f in $\mathbb{R}^n$ with its integrals over lines is usually called the *X-ray transform* of f.[15] Further important generalizations of these results to integral geometry were obtained by S. Helgason and A. M. Cormack in the 1960s.

For odd dimension n, the reconstruction of a function f in $\mathbb{R}^n$ from its integrals over hyperplanes is local in the sense that f(z) is determined by the integrals along the hyperplanes through a neighbourhood of z. For the case of even dimension n, this is not true, but integrals over *all* hyperplanes are needed. Thus, for example, in Radon's original formula (7) for $n = 2$, all line integrals of f are required. Various computer algorithms have been developed to meet the computational demands of the inversion formulas with incomplete or limited data.

Radon worked in pure mathematics without specific applications in mind. The first radiological applications of inversion methods were developed in 1960s by G. Hounsfield and A. M. Cormack. They received the 1979 Nobel prize of medicine for the introduction of computerized tomography.[16] This technique in different variations has become the standard tool of detecting cancer tumours in diagnostic medicine.

The basic physical model is the following. Assume that the density of a body B at each of its points can be expressed by a function f(z) in $\mathbb{R}^2$. When an X-ray with initial intensity I_0 travels through the body B along line L, its intensity is attenuated at the point z by the relative loss $\Delta I/I = f(z)\Delta z$. Let I_1 be the intensity of the ray after having passed the body B. Then it follows from this linear model that

$$I_1 = I_0 \exp\left(-\int_L f(z)dz\right)$$

so that

(8) $\int_L f(z)dz = \ln (I_0/I_1)$.

As the values of I_0 and I_1 are physically measurable for each L, formula (8) in effect gives us the line integrals of f over L. By putting these measured values in Radon's formula (7), the density f(x) of body x can be computed.[17]

---

[14]Thus, in $\mathbb{R}^3$ a hyperplane is an ordinary plane which divides $\mathbb{R}^3$ in two half-spaces.

[15]Modern proofs of the generalizations of formula (7) employ sophisticated tools from Fourier analysis. The "Fourier slice theorem" gives a simple mathematical relation between the Fourier transform f^ of a function f: $\mathbb{R}^n \rightarrow \mathbb{R}$ and the Fourier transform of the Radon transform R(f) of f (Natterer 2001, 11). This implies that f itself can be inversely calculated from the values of R(f). Similar results holds for f^ and the Fourier transform of the X-ray transform of f.

[16]Greek *tomos* means 'slice', 'section', or 'cutting'.

[17]Note that model (8) can be understood as solving an estimation problem of the form (3): I_0 is the input $\underline{x}$, I_1 is the output y, and the infinite number of values of function f(z) correspond to the unknown parameters $\underline{a}$.

In practice it is not possible to process an infinite number of lines L, so that inverse problems in tomography are ill-posed. Two scanning geometries are used (Natterer 2001,2): *parallel scanning* employs a single source and a single detector, and a set of equally spaced parallel lines are taken for a finite number of equally distributed directions; in *fan-beam scanning*, the source runs a circle around the body and fires a fan of X-rays recorded by an array of detectors. The task of tomography is then to reconstruct a two-dimensional computer image of the body B by using inversion algorithms. For three-dimensional images, the body may be scanned layer by layer, or a spinning source may fire a cone of X-rays.

In some "local" applications of tomography, only a part of a body is investigated. In single particle emission computerized tomography (SPECT), the radiation source is inside the body. In positron emission tomography (PET), detectors measure coincidences of two particles ejected by the sources in opposite directions. In impedance tomography, X-rays are replaced by flows of electricity through a body. In diffraction tomography, measurements concern refractions of beams and ultrasounds from scattering objects. In the limiting case, an opaque implant is present. Similar mathematical methods are in used in electron microscopes, radio telescopes, image reconstruction, acoustics, and radar theory.

Today there are still many inverse problems which wait for their theoretical solutions in mathematics, statistics, and computer science. Applications of these problems can be found physics, chemistry, astronomy, geophysics, engineering, environmental studies, biology, clinical medicine, archaeology, cultural anthropology, neurology, psychology, and cognitive science.

The growing success of inverse problems as a branch of applied mathematics reconfirms Peirce's great insight of the significance of abductive inference in science. For friends of abduction, this observation provides a rich source of case studies from various scientific fields. It also gives a strong argument against those anti-realist philosophers who have denied that abduction has a legitimate role in science.[18]

It is especially interesting to see that methods of solving inverse problems serve at the same time both discovery and justification – two functions that philosophers of science usually separate in their discussions of abduction (cf. Chap. 1). For example, algorithms in medical imaging at the same time detect tumours and give well-supported descriptions of their geometrical shape. The example of tomography thus gives an argument against the view that restricts the role of abduction solely to the context of discovery.

We have seen that the historically fundamental theorem, proved by Radon already in 1917, is the strongest kind of mathematical result about inverse reasoning: it guarantees the existence of a unique backward solution, and gives an explicit formula for computing the solution. In practical applications in diagnostic medicine and other areas of science, the theoretical framework has to be modified by allowing

[18]The critics of realism have the burden of proof in showing why abduction may be employed in some areas of science but not in others (see Boyd 1984, 67).

incomplete data and approximate statistical methods. This reasoning is abductive in Peirce's original sense, since it proceeds from effects to causes and the computed solution gives the best explanation of the measured data (cf. (4)). The process of constructing computer images of cancer tumours in tomography resembles human perception of external physical bodies, which Peirce regarded as a limiting and compelling case of abduction.[19] Indeed, the computerized inverse algorithms in diagnostic medicine are so reliable that we are ready to risk our health, even urgent matters of life and death, on their routine applications.

4.4 Abduction and Biological Evolution

With influences from the Harvard fellow Chauncey Wright, Peirce was a great admirer of the theory of biological evolution. The Darwinian controversy was in large part "a question of logic", he stated in 1877 (*CP* 5.364). According to Peirce, Darwin applied the statistical method to biology in same way as Clausius and Maxwell to the theory of gases:

> In like manner, Darwin, while unable to say what the operation of variation and natural selection in any individual case will be, demonstrates that in the long run they will, or would, adapt animals to their circumstances.

In this way, Darwin's theory is able to give a probabilistic explanation of many facts about the development of organisms. But Peirce himself did not give examples from the study of biological evolution to illustrate his notion of abduction.

Darwin's theory of evolution involves abduction in three ways. First, evidence for this theory is largely abductive, based upon its explanatory power.[20] *The Origin of Species* (1859) presents the theory of gradual non-directed evolution by the mechanisms of variation and natural selection which guarantee that the fittest organisms

---

[19]In human perception, visual representations are transformed to mental pictures of three-dimensional objects. It is beyond the scope of this work to study the question, whether the human mind employs in perception processes that are similar to some inverse algorithms. But at least it is interesting to observe that the idea of looking at an external object from different directions or perspectives, so prominent in Radon's theorem (11) which involves all possible lines passing through the body, has since the early twentieth century been a key element of Husserl's phenomenology of perception (see Husserl 1913), the psychology of perception (see Kaila 1960), and the cubist art of Braque and Picasso (see Hintikka 1975). It is known that the geometrical shape of a three-dimensional opaque and convex object can be reconstructed, if we can observe it (or its projections, shadows, and scattering patterns) from all directions. For example, the spherical form of an asteroid can be inversely calculated from radar data. The abstract mathematical problem was proposed by Hermann Minkowski already in 1903: the shape of a convex surface uniquely determines its Gaussian curvature, and inversely the curvature determines the shape of the surface.

[20]While the theory of evolution is not able to predict the actual course of evolutionary paths, which depend on random mutations, it has some predictive power about the adaptation of species to their environment. See Williams (1982). Similarly, geologists can explain earthquakes when they have occurred, but their ability to predict them is limited.

will survive in the struggle for existence. Darwin appeals to John Herschel's *vera causa* principle by pointing out the analogy between natural selection and domestication (i.e. plant and animal breeding). In harmony with William Whewell's principle of unification, or "consilience of inductions" (cf. Sect. 6.4), Darwin then shows that his theory is able to explain a variety of phenomena: instincts, the fossil record, the geographical distribution of present species, and morphological and embryological facts.[21] Darwin's theory of natural evolution gives a better explanation of these phenomena than creationism or Intelligent Design theories.[22]

Secondly, many current traits of organisms can be explained as adaptations, i.e. by the assumption that at some earlier time there was a selection or fitness advantage for them. This inference to an explanation is again abductive. But adaptationism has a bad reputation due to problematic applications: the past selection situation is often reconstructed in a careless way, and the assumed genetic base is not put into independent tests. Further, what once were adaptations need not be adaptive any more (see Sober 1984, 1993).

Thirdly, on the basis of Darwinian evolution theory and palaeontology, the history of life on earth can be represented by a tree with a common descent and branches corresponding to speciation. The current version of *the tree of life* in 3,5 billion years includes 2,3 million species with names,[23] and there may still be missing links and unidentified species. Evolution is a process which goes forward in time, and the reconstruction of the evolutionary trees on the basis of present evidence (contemporary forms of life and fossil records of extinct species) is backward inference or a retroductive task in Peirce's sense.[24]

The method for constructing such evolutionary trees in biological phylogenetics is known as *cladistics*, whose main ideas were established by the German entomologist Willi Hennig in 1950. The classical approach is to measure the similarity of species by the (possibly weighted) number of their common traits or characters, but restricted to the "derived" or novel features of a species and its descendants (see Sober 1988). The rival approach of *pheneticism* is to apply numerical methods by

[21]For the structure of Darwin's argument, see Ruse (1982). Cf. Kitcher (1981) and Morrison (2000).

[22]See Sober (1993, 2008) and Sarkar (2007).

[23]See *tree.opentreeoflife.org*.

[24]The connection between evolutionary trees and Peirce's abduction was noted in Niiniluoto (1985). Sober (1988) gives a detailed analysis of phylogenetic reasoning in biology as abduction or "a kind of inference to the best explanation". For Sober, this form of abduction infers a common ancestry by the principle of common cause, but he is cautious in applying this principle in biological contexts, since in some cases (e.g. homoplasies) separate causes explanations should be preferred. Schurz (2008a, 2017) mentions evolutionary reasoning as an example of common cause abduction. Kirk Fitzhugh, curator at the Natural History Museum of LA County, has argued that phylogenetic hypotheses are abductively generated as answers to contrastive explanatory questions (see Fitzhugh 2006, 2007). Nepomuceno (2014) notes briefly that the study of the ancestors of *Homo sapiens* involves abduction. Quinn (2016b) argues that the discovery of patterns of descent in biological systematics is based on inference to the best explanation. Her answer to van Fraassen's bad lot argument (cf. Sect. 7.2) is to include all possible historical stories or trees as alternatives, but this does not appreciate the computational limitations of phylogenetic analysis (cf. footnote 32).

measuring the overall similarity of species with respect to all of their traits (see Sneath and Sokal 1973).[25] Then a "cladogram" or a "dendrogram" is constructed so that more similar species have more recent ancestors. This method follows a principle of parsimony or simplicity with respect to gradual evolution: the process of speciation flows with the minimum number or length of evolutionary steps. In the construction of cladograms, special attention has to be given to the problem of convergence, i.e. "homoplasies" or accidental similarity of features in different phyletic lines without common ancestors (e.g. the wings of bats, birds, and insects). These are contrasted to "homologies" which indicate common ancestry (e.g. the wings of sparrows and robins). The method of maximum parsimony minimizes assumed homoplasies (see Sober 2015). Another problem is reticulation, i.e. the possible fusion of lineages (e.g. hybridization of plants and micro-organisms). A typical application of cladistics is to argue that birds are the closest existing relatives to the Saurischian or "lizard-hipped" dinosaurs (see Fig. 4.3).[26]

In plant and animal taxonomy, cladograms are usually constructed so that OTU's (Operational Taxonomic Units) are placed in the end leafs of bifurcating trees, while internal nodes are left to HTU's (Hypothetical Taxonomic Units). By using dating methods such cladograms can be completed with a temporal dimension, so that a *phylogenetic tree* represents the evolution of current species from earlier (possibly extinct) species by ancestral-descendant relationships. Such evolutionary trees for hominins, which are rapidly changing with new fossil findings, take into account feature similarities together with evidence about the historical migration and interaction of the members of the species (see Fig. 4.4).

The methods of feature matching in numerical taxonomy can be applied on the level of the phenotype (observable properties) or genotype (genetic structure). Today one can distinguish species trees and gene trees, which need not be congruent. The classical methods of cladistics have been complemented by new computational approaches, and the criterion of parsimony is contrasted with probabilistic maximum likelihood and Bayesian methods.[27]

[25]Hull (1988) tells a lively story about cladistics and pheneticism as rival research programmes.

[26]This was the main theme of an exhibition in the British Museum (Natural History) in 1979 (see Hedley 1979). Today many taxonomists are ready to assert that birds *are* dinosaurs.

[27]Joe Felsenstein showed in 1978 that maximum parsimony is not statistically consistent, i.e. it need not converge on the correct tree with increasing data. The methods of maximum parsimony and maximum likelihood, including their different assumptions about the biological evolutionary process, are compared by Sober (2008), 332–352. Recent approaches of cladistics are summarized by Quinn (2017).

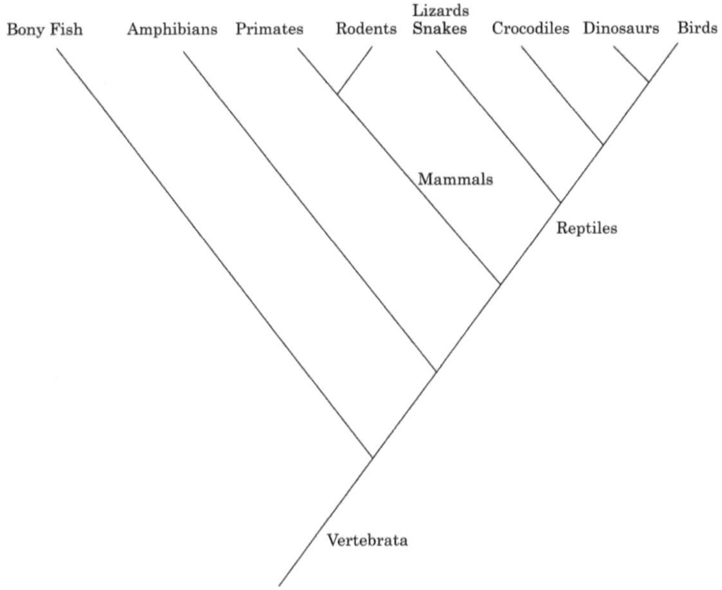

Fig. 4.3 A cladogram for vertebrata

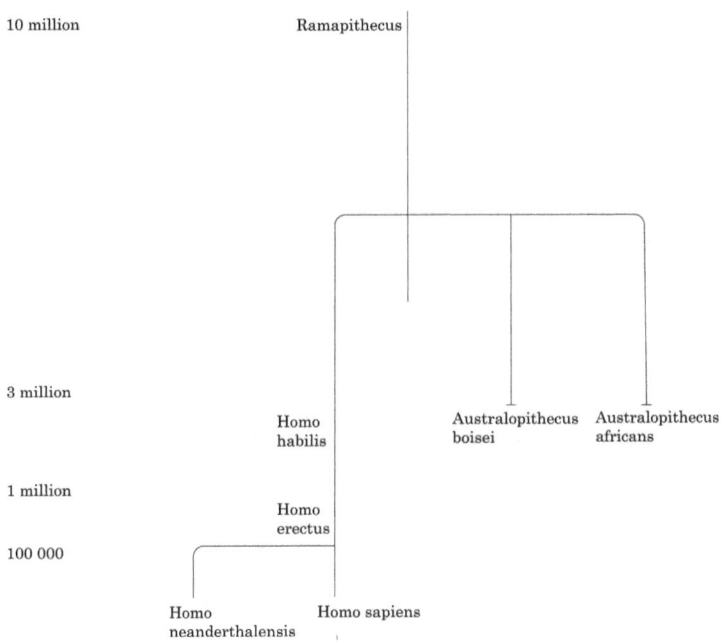

Fig. 4.4 Richard Leakey's phylogenetic tree for hominids (in 1977)

4.5 Evolutionism in the Humanities

Starting from the 1860s, the idea of evolution was applied in the study of culture: sociology (Spencer), philology (Schleicher), ethnology (Bastian), anthropology (Tylor, Morgan), folkloristics (Krohn), epistemology (Peirce), and ethics (Westermarck). Today there are research programmes for reducing cultural evolution to biology (evolutionary psychology, sociobiology) or showing it to be analogous to biological evolution (memes as "cultural genes"), but the success of such attempts is limited by the fact that *cultural evolution* is more "Lamarckian" than Darwinian: variation is not random (cultural entities are intentionally produced with a goal) and transmission involves inheritance of acquired characteristics (by learning and symbolic languages).[28] Moreover, fitness is measured by cultural utilities or social values rather than by success in reproduction.[29] In spite of these differences, evolutionism in the humanities has led to abductive inferential patters which are similar to cladistics in biological taxonomy.

In Peirce's example of Napoleon, the physical objects constituting the documents and monuments about this emperor have not changed over time (possibly apart from slight erosion). But in his discussion about the evolution of science, in the manuscript "Lessons from the History of Science" in 1896, Peirce considered the possibility that judgments are fortuitously modified at every recall to the mind or opinions are perpetually enriched by new observations.[30] Such gradual changes are typical in cultural evolution. For example, Schleicher assumed that, while languages create new words or adopt them as cultural loans, they gradually develop toward greater phonetic and structural simplicity. He concluded that the evolution of languages can be represented by family trees which grow from a basic or original form to several branches. Thus, for example, the tree of Finno-Ugric languages starts from an assumed proto-language and branches to existing languages like Finnish, Estonian, Hungarian, Sami languages in Lapland and endangered minority languages in the present territory of Russia (see Fig. 4.5).

Cultural items in the oral tradition are changed when they are transmitted from one generation to another or from one village to another. In Finland, Elias Lönnrot

[28]When "Lamarckism" is associated with the inheritance of acquired characteristics (Ruse 1982, 7), one may recall that Darwin's own (pre-Mendelian) pangenesis hypothesis in 1868 allowed that the use and disuse of organs may lead to structural modifications in cells. One may also doubt whether Lamarck himself attributed voluntary or intentional action to animals (see Burkhardt 2013).

[29]Following Charles R. Lumsden and Edward O. Wilson, one may also speak about "gene-culture coevolution", or interaction between biological and cultural evolution by means of inherited "epigenetic rules": some adaptions (e.g. speech) may confer dispositions to learn a new disposition under the influence of environmental factors (e.g. mastery of a specific natural language like Chinese or Finnish), and the behavioral exercise of these cultural capacities may lead by natural or sexual selection to different rates in reproduction and survival. See Fetzer (1985).

[30]According to Peirce, the usual process of scientific evolution is "cataclysmal" or "by leaps", but it may also be gradual in parallel to the Darwinian and Lamarckian modes. See Peirce (*CP* 1.103–109); Niiniluoto (1984), Ch. 4.

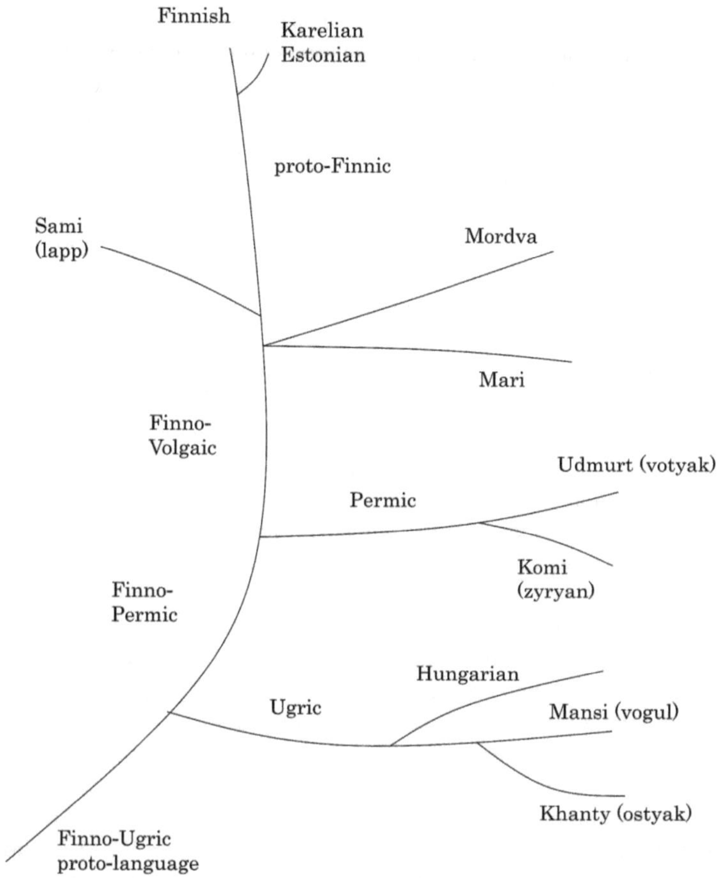

Fig. 4.5 Paavo Ravila's language tree for Finno-Ugric languages (in 1946)

collected the rich oral traditions of epic in *Kalevala* (1839/1849) and lyric in *Kanteletar* (1840). Inspired by Lönnrot, Julius Krohn and his son Kaarle Krohn, professors at the University of Helsinki, developed a systematic way of finding the original form of a folk song, tale, myth, legend, joke, or proverb. This historic-geographical approach was called "the Finnish research method" in folkloristics (see Krohn 1926).

Before the age of electronic Xerox copies, written texts (e.g. manuscripts, poems) were modified when they were copied by hand: hand-made copies may include errors which then repeat themselves in later copies. The method of identifying the original text and showing the historical development of its variants is called *textual criticism*. As its task is to construct a "stemma" which places all these variants as nodes of a tree, textual criticism is also called *stemmatics* or *stemmatology*.

Given the evolutionary influences of the human and social sciences, it is natural to expect that the inferential structure of textual criticism and stemmatology in cultural sciences is similar to cladistics in biological taxonomy.[31] Platnick and Cameron (1977) argued that "cladistic analysis is a general comparative method applicable to all studies of historical interrelationships based on ancestor-descendant sequences", and illustrated this thesis by analyzing the inferential structure of textual criticism and linguistic reconstruction, based on detected homologies between texts or languages. In particular, for languages such similarities may concern vocabulary and grammar.

New confirmation for the thesis was given by Peter Robinson and Robert O'Hara (1992), who applied the software package PAUP, developed originally for phylogenetic analysis, as a computer method in stemmatology. Their method is based on the maximum parsimony criterion for ranking alternative trees. Here we give some examples of further studies in the field of digital humanities.

Russell Gray, director of the Department of Linguistic and Cultural Evolution at the Max Planck Institute for the Science of Human History in Jena, has since 2005 applied phylogenetic methods in the study of the origin and prehistory of Indo-European languages.

Today computer-assisted stemmatology can achieve in a few seconds results which took several weeks for a humanist scholar with paper work. Digital methods are also more reliable than the old manual ones. The historian Tuomas Heikkilä and computer scientist Teemu Roos at the University of Helsinki have compared alternative algorithms for an artificial data set *Heinrichi* with 1200 words in 67 variants and a real-life data set *The Legend of St. Henry* with fifty medieval versions (see Roos and Heikkilä 2009). For the purpose of comparison, they define the distance between stemmata as the average of the index values u(A, B, C) for all triples of distinct nodes A, B, C with available texts, where u(A, B, C) takes values 1, ½ or 0 depending on whether the proposed stemma agrees with the correct one about which one of B and C is closer to A. They propose their own RHM method which measures simplicity with an approximation of Kolmogorov complexity by a compression program gzip, and show that it performs better or at least as well as PAUP. The RHM method restricts its attention to bifurcating trees, where each interior node has exactly three neighbours, but still the number of such trees is exponentially large.[32] Existing texts are placed in the leaf nodes, whereas the internal nodes are reserved for missing variants. It is recognized that the tree structure may be upset by the problem of contamination, which arises when a text is copied from two sources. This corresponds to reticulation in the biological case. The result of applying RHM to the Legend of St. Henry is shown in Fig. 4.6.

The cultural anthropologist Jamshid Tehrani, together with computer scientists Quan Nguyen and Teemu Roos, has studied the interaction of oral and literary

[31]For this observation, see Niiniluoto (1985), who also argues that textual criticism and cladistics use abductive inference. Sober (1988), 6, who notes that phylogenetic inference and historical studies of languages and ancient texts are instances of genealogical inference, refers to an earlier article by Platnick and Cameron (1977).

[32]For example, the number of alternative bifurcating trees for 56 terminal taxa is of the order 10^{74}.

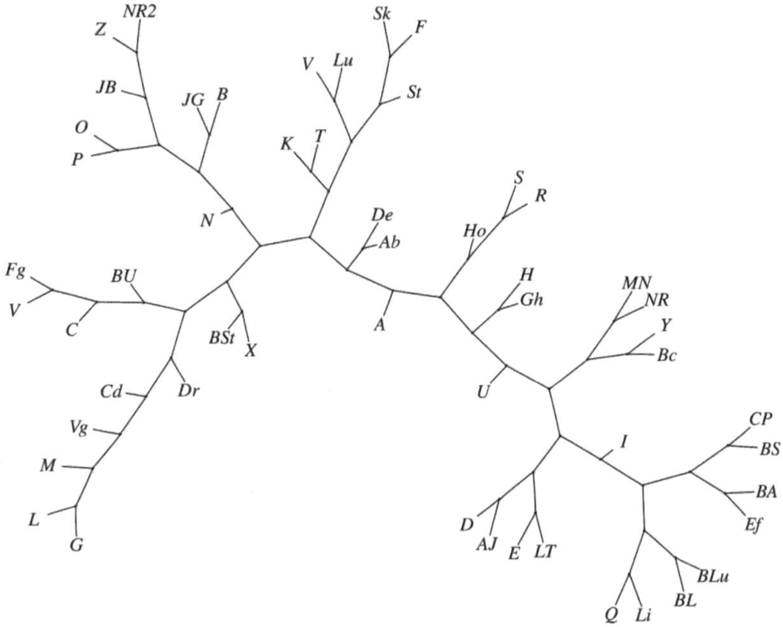

Fig. 4.6 The Legend of St. Henry (by RHM)

traditions by comparing the early literary versions of Little Red Riding Hood (Perrault in 1687 and the Brothers Grimm 1813) with twenty-two other literary or oral versions of the tale (among them an eleventh-century Latin poem and several Franco-Italian oral versions collected in the nineteenth century). Their aim is to settle the old controversy in fairy tale scholarship: did the oral tradition of this famous tale derive from literary versions or vice versa? They apply the maximum parsimony method PAUP, two network approaches NeighbourNet and TRex which allow the addition of reticulation edges to the tree, and their own likelihood-based method PhyloDAG. Following the Jukes-Cantor model of probabilistic sequence evolution in biology (cf. Sober 2008, 337), they propose a probabilistic model for lineal (ancestor – descendent) relations, place some extant taxa at internal nodes, and add new edges to obtain maximum likelihood of the network. The results derived from PhyloDAG strongly favour the oral origins hypothesis (see Fig. 4.7).

4.6 Conclusions

We have argued in Sect. 4.4 that the reconstruction of the tree of life by means of cladistics analysis is a retroductive task in Peirce's sense. Its premises include the Darwinian theory of evolution with data about extant and extinct species. Its conclusion, a phylogenetic tree, is a much more complex statement than a singular

Fig. 4.7 Little red riding hood

claim about a particular individual like Napoleon Bonaparte, since it tells about the historical development of species by bifurcating lineages.

We have also shown in Sect. 4.5 that historical linguistics, literature studies, and folkloristics apply similar methods in the construction of family trees of languages and tree-like stemmas of written or oral texts. Cladistics, textual criticism and stemmatology turn out to be closely related, so that similar (and even same) computer algorithms can be used in biology and cultural studies. Active investigation of these computational methods may reveal some interesting differences in their details and conditions of application. But for us it is most important to note that these abductive methods serve both discovery (heuristic selection of one of the trees among numerous alternatives) and justification (argument that this tree is a plausible or likely reconstruction of the real historical evolution).

These observations, together with results from Sect. 4.3, support Peirce's thesis that abduction, both as a method of discovery and justification, is an important form of reasoning in many different scientific disciplines.

Chapter 5
Abduction as Discovery and Pursuit

Abstract The Scientific Revolution, which replaced ancient doctrines in the early modern age, convinced researchers that science is a systematic way of seeking *new* knowledge. For this reason, the method of scientific inquiry has to leave room for the heuristic generation of novel ideas. Many philosophers have sought a pattern of discovery in inductive reasoning, but the champions of the hypothetico-deductive model in science altogether denied the possibility of a logic of discovery (Sect. 5.1). Charles S. Peirce argued that "all the ideas of science come to it by way of abduction", and described abduction as a kind of "insight". In Sect. 5.2, we evaluate the potential of abduction as a logical reconstruction of the process of discovery by considering the ideas of Norwood Russell Hanson, Herbert Simon, and other "friends of discovery". In Sect. 5.3, we discuss Larry Laudan's alternative interpretation that abduction gives reasons for the pursuit of scientific hypotheses. In this sense, abduction gives before-trial criteria for identifying plausible or testworthy hypotheses. The GW-model is also related to the idea that inverse abductive reasoning is relevant for pragmatic human action. We give reasons to doubt that the pursuit and acceptance of hypotheses can always be sharply separated from each other.

The Scientific Revolution, which replaced ancient doctrines in the early modern age, convinced researchers that science is a systematic way of seeking *new* knowledge. For this reason, the method of scientific inquiry has to leave room for the heuristic generation of novel ideas. Many philosophers have sought a pattern of discovery in inductive reasoning, but the champions of the hypothetico-deductive model in science altogether denied the possibility of a logic of discovery (see Sect. 5.1). Charles S. Peirce argued that "all the ideas of science come to it by way of abduction" (*CP* 5.145). In Sect. 5.2, we evaluate the potential of abduction as a logical reconstruction of the process of discovery. The alternative interpretation, that abduction gives reasons for the pursuit of scientific hypotheses, is discussed in Sect. 5.3.

I. Niiniluoto, *Truth-Seeking by Abduction*, Synthese Library 400,
https://doi.org/10.1007/978-3-319-99157-3_5

5.1 Patterns of Heuristic Reasoning

In ancient Greece, philosophy of nature started as an attempt to understand the permanent or invariant structure of reality by means of human reason, independently of religions and old myths (cf. Kaila 2014). In Plato's dialogue *Theaetetus*, Socrates considers the analysis of knowledge (Gr. *episteme*) as "true belief with *logos*", but in spite of his reservations this proposal has been consolidated as the "classical definition" of knowledge as justified true belief. Aristotle formulated the logical principles of supplementing factual *knowledge that* with demonstrated *knowledge why* (cf. Sect. 2.1). The Aristotelian conception became the dominant medieval view of *scientia* as a contemplative state of knowledge. It was only in the early seventeenth century, after the Scientific Revolution of Nicolaus Copernicus, Johannes Kepler, and Galileo Galilei, that philosophers like the empiricist Francis Bacon and the rationalist René Descartes convinced their contemporaries of the need to emphasize the search of *new* truths in science.

The generation of new knowledge in science involves both discovery and invention. To discover is *heureka* in Greek and *invenire* in Latin. So typically a scientist discovers new facts and regularities in nature by inventing symbolic statements, laws, and theories to express their findings. While psychologists may empirically study the mental processes of discovery, viz. psychological heuristic, the dream of many methodologists has to been to find a logical approach to discovery, viz. *logical heuristic*. As we have seen in Sect. 2.2, such a systematic pattern of heuristic inference was formulated already in Greek geometry as the method of analysis and synthesis. Medieval and Renaissance philosophers, who developed the regressive method of resoultion and composition, appealed to Aristotle's deductive syllogisms or his account induction (*epagoge*) as the method of reaching the first premises of science (*An. Priora* II, 23).

The idea of a mechanical system of reasoning was upheld by some medieval scholars. Raymond Lull in his *Ars Magna* (1305) tried to construct a system of solving inferential problems with moving circles, trees, and symbols. It did not work at all, but the idea was continued by G. W. Leibniz's program of logic as a *calculus ratiocinator*. Its modern realization in the mathematical logic of Boole, Frege, Russell, Peirce, and Hilbert led in the late nineteenth century and early twentieth century to sophisticated and powerful formal systems, which can be used to the check the validity of given arguments. Logical rules of inference are permissive in telling what inferences are allowed, but do not guide which of the potentially infinite consequences we should draw. In Hintikka's (1998) terms, they are definitory rules of logical validity, not strategic rules for seeking knowledge (cf. Sect. 3.5). Moreover, already first-order logic with relations is undecidable, so that there is no mechanical test for validity. In this sense, modern systems of formal logic are not "logics of discovery".[1] This seems to confirm the view that the generation of new ideas should follow some ampliative (content increasing) method of reasoning.

The *inductivist* view of science identified enumerative inductive generalization as the method of discovery. In spite of the doubts by the sceptic Sextus Empiricus (second century), Francis Bacon, and David Hume, this view gained popularity among the eighteenth-century Newtonians and the nineteenth century positivists (like August Comte). Eliminative versions of induction were formulated by Bacon and John Stuart Mill. In his *A System of Logic* (1843), Mill defined induction as "the operation of discovering and proving general propositions".

Bernard Bolzano's *Wissenschaftslehre* (1837) includes a chapter on the "art of discovery" (Erfindungskunst). In the English translation, the title of the chapter is "Heuretic". It discusses, among other things, "the discovery of causes for given effects" (see Bolzano 1972). His examples of such discovery of causes resemble Mill's rules of reasoning (Methods of Difference, Agreement, and Concomitant Variation).

A well-argued and careful attack on inductivism was given by William Whewell in his *The Philosophy of the Inductive Sciences* (1840) and *On the Philosophy of Discovery* (1860). In spite of his acceptance of Bacon's recommendation for "a continuous advance from observation, by limited steps, through successive gradations of generality" (Whewell 1860, 131), he argued that every step in scientific discovery involves the suggestion and appropriation of a new conception which "colligates" the known facts. Like Aristotle and Kant, Whewell thus linked his "induction" to concept formation, which is absent in ordinary inductive generalization. Hence, science does not advance in the way suggested by Bacon, i.e. by accumulating a great collection of facts from which philosophers might then extract scientific truths by the application of a right method (*ibid*., 153). Such methods of procedure "do not belong to the *Epoch of Discovery*, but to the Period of *verification* and *application* of discovery which follows" (*ibid*., 154).

Whewell added to his "discoverers' induction" (Snyder 1997) precise criteria for good hypotheses. His studies were thus influential in the development of the hypothetico-deductive (HD) model of scientific inference. But he also argued that Bacon did not "justly appreciate the sagacity, the inventive genius, which all discovery requires" (*ibid*., 138). Such sagacity leads to "happy guesses" by "framing several tentative hypotheses and selecting the right one" (Whewell 1847, 467–468).

Peirce knew well Whewell's work and took sides with him against Mill's inductivism (*EP* 2:46). For Whewell, induction involves discovery, since the given facts are first discovered to satisfy some new conception π, and only then this "superinduced" concept is used as a basis of inductive generalization. Peirce, on the other hand, argued that in a valid induction the character π has to be "predesignated", rather than "suggested by the examination of the sample itself", so that "induction can never make a first suggestion" (*RLT*, 139). In this sense,

[1] Chapter 2 discusses some formal systems, inspired by Peirce's notion of abduction, which attempt to serve as "logics of discovery" – or "logics of generation and evaluation" (Aliseda 2006).

Fig. 5.1 Pascal's triangle

				1				
			1		1			
		1		2		1		
	1		3		3		1	
1		4		6		4		1
…			…			…		…

Peirce's conception of abduction resembles Whewell's induction, since it requires imagination (*CP* 1.46), while the element of discovery is lacking in inductive generalization.

That induction may still have some value in mathematical discovery has been argued by Georg Polya (1957). For example, there are mathematical theorems valid in the two-dimensional real space $\mathbb{R}^2$, which can be generalized to $\mathbb{R}^n$ for arbitrary natural number n. The sums of the numbers in the rows of Pascal's triangle (see Fig. 5.1) are 1, 2, 4, 8, 16, … and one can easily observe that they are successive potencies of 2, i.e. of the form 2^n, $n = 0, 1, 2, \ldots$

This generalization can be tested by calculating the sum on the next row to be 32 or 2^5. In the next stage the mathematician can prove the generalization deductively by using the so called principle of mathematical induction. Polya (1957), 117, concludes:

> Many mathematical results were found by induction first and proved later. Mathematics presented with rigor is a systematic deductive science but mathematics in the making is an experimental inductive science.

But one should note that the discovery of the general form 2^n is a colligation in Whewell's sense rather than a simple inductive generalization. Polya also recognized the importance of "plausible" reasoning in problem-solving, which goes inversely from consequences to their antecedents.[2]

Mathematicians also use their power of *imagination* to discover new theorems which are proved only much later. Famous examples include Fermat's Last Theorem in 1637, proved by Andrew Wiles in 1994, and De Morgan's Four Colour Theorem in 1852, proved by K. Appel and W. Hagen with the help of computers in 1976. This is sometimes called "intuition", but Peirce contrasted the scientific method with the unreliable methods of tenacity, authority, and intuition so that he preferred the term "insight" for sudden unexpected discoveries (see Shanahan 1986; Paavola 2005): "The abductive suggestion comes to us like a flash. It is an act of *insight*, although extremely fallible insight" (*CP* 5.181). Working scientists like Henri Poincaré have reported such discoveries in dreams, city crowds, and trains.

Psychological studies of the processes of discovery have given explanations which suggest that insight is not incompatible with idea that discovery has logical

[2]Lakatos (1976) argued that, instead of induction, discovery in mathematics follows Popper's methodology of conjectures and refutations.

and inferential aspects. With inspiration from Poincaré, the mathematician Jacques Hadamard (1945) has distinguished four stages in invention. First, in the "preparation" stage, the problem is formulated and a solution sought. If this fails, then in the second stage of "incubation" our unconsciousness goes on work on the problem by combining ideas in new ways and selecting the most beautiful or promising ones. In the third stage of "illumination", a solution suddenly appears in our conscious mind. In the fourth stage, we still verify or precise the solution. Hadamard's account anticipates the psychological theory of problem-solving by Herbert Simon (1977), who argues that human thinking consists of "elementary information processes" which are in principle similar in conscious and subconscious levels. What is more, similar patterns of reasoning are followed by systems of Artificial Intelligence, such as chess-playing programs, which must use "highly selective methods of trial and error" in "heuristic search".

Sometimes discoveries need good luck. Following Horace Walpole's suggestion in 1754, based on the Persian fairy tale *The Three Princes of Serendip*, such unexpected discoveries, where valuable things are found without being searching for them, are said to be based on *serendipity* (see Roberts 1989). Famous cases include the discovery of X-rays by Wilhelm Röntgen in 1895 and the penicillin by Alexander Fleming in 1928. But such good luck is usually associated with intensive research activity which helps to recognize the importance of the new finding: as Louis Pasteur stated, "chance favors the prepared mind". As shown by historical cases of simultaneous discoveries (e.g. differential calculus by Newton and Leibniz, conservation of energy by Carnot and Joule), one should not exaggerate the role of luck in discovery.

The *HD model* starts from criticism of inductivism and concludes that there is no inference to scientific hypotheses. Carl G. Hempel (1966), 15, summarizes this position as follows:

> The transition from data to theory requires creative imagination. Scientific hypotheses and theories are not *derived* from observed facts, but *invented* in order to account for them.

Karl Popper made a sharp distinction between "the process of conceiving a new idea, and the methods and results of examining it logically":

> There is no such thing as a logical method of having new ideas, or a logical reconstruction of this process. My view may be expressed by saying that every discovery contains 'an irrational element', or 'a creative intuition', in Bergson's sense (Popper 1959, 31–32).

This quotation is from the English translation of Popper's 1934 work *Logik der Forschung*, misleadingly entitled *The Logic of Scientific Discovery*.[3]

A weak point about the HD model is it that leaves the discovery of scientific hypotheses as a mystery, which may invite various kinds of speculative accounts. For example, Pierre Duhem's (in 1907) as such attractive comparison of a physicist to a flower in the field who just "keeps its corolla open to the insect carrying the generative dust of the fruit" (Duhem 1954, 256) has an implicit religious message,

[3]In his later works, Popper was concerned with the growth of knowledge (see Popper 1963, 1972).

which is made explicit in Abbé Gratry's claim, mentioned by Peirce (see *CP* 2.689), that every true induction is a "miracle" or a result of "an immediate inspiration from on high".

Peter Achinstein (1971), 142–149, gives Gay-Lussac's Law in 1809 as an example of inductive discovery. It was based on the observation that water is composed of 100 parts by volume of oxygen and 200 of hydrogen, but the generalization that other gases combine in simple ratios also used "a good deal of theoretical and experimental background information". Achinstein further argues that Joseph Priestley's 1767 discovery of the inverse-square law of electrostatics was based on analogical inference from Newton's law of gravitation which has the same mathematical form (*ibid.*, 153–155).

Isaac Newton is sometimes accused of inductivism, since he refused to "feign hypotheses" and stated that in analysis general conclusions from experiments and observations are drawn "by induction". But the analytic method, which involves consideration of interdependencies of factors in experimental situations, is much more complex than any rule of inductive generalization (see Hintikka and Remes 1974; cf. Sect. 2.2). Newton's famous statement that he "deduced" the form of his law of gravitation from "phenomena", i.e. Kepler's laws, is not always taken seriously, but this inference can be reconstructed as a genuine deductive argument (see Harper 2011).

These historical examples serve as an antidote to the HD thesis that there is no inference to theoretical hypotheses. At least in special cases, induction, analogy, and even deduction may play a role in the generation of theories. But we now turn to Peirce's suggestions about abduction which give much richer possibilities for the analysis of heuristic reasoning in science.

5.2 Abduction as Discovery

Peirce repeatedly argued that "all the ideas of science come to it by way of abduction" (*CP* 5.145). Elsewhere he stated that science "advances by leaps", and "the impulse for each leap is either some new observational resource, or some novel way of reasoning about the observations" (*CP* 1.109). There is no incongruity here, when we remember that perception is for Peirce a limiting case of abduction (see Sect. 4.1).[4]

[4]Peirce also stated in 1883 that the scientific task of "discovering laws" is accomplished by induction, and "discovering causes" by hypothetic inference (*CP* 2.713). In this context, he is speaking about induction as statistical generalization (see (1.15)), where the reasoning from the sample to a population M presupposes that the attribute P to be detected in the sample should be "predesignated" (*CP* 2.736). Induction in this sense presupposes a reference class M and an initial choice of a concept P, and it only starts after the form of the statistical law "The proportion x of the Ms are P" has been agreed upon. Induction amounts to the estimation of the unknown population parameter x on the basis of a subsequently drawn sample from M. The result of such inference –

Serious discussion of abduction as a logic of discovery started with Norwood Russell Hanson in *Patterns of Discovery* (1958). Hanson belonged to the generation of new philosophers of science (among Thomas Kuhn, Karl Popper, Stephen Toulmin, and Paul Feyerabend) who thought that logical empiricists have given a too static picture of scientific theories as "finished research reports". The focus, he suggested, should be in "theory-finding" instead of theory-testing. Hanson rejects both inductivism and HD accounts by going through what Peirce called "the greatest piece of retroductive reasoning ever performed" (*CP* 1.71-74): Kepler's discovery of the elliptic orbit of Mars in *De Motibus Stellae Martis*. According to Peirce, Kepler at each stage of his inquiry had "a theory which is approximately true" and proceeded to "modify this theory in such a way to render it more rational or closer to the observed fact". Hanson elaborates Peirce's reconstruction of Kepler's inference, which used Tycho's observations as data, started from the assumption that the sun must have something to do with causing the planets to move in their orbits, rejected the traditional thesis of circular orbits, tried several alternatives but rejected them because of their too large errors, and finally arrived at the conclusion that the path of Mars is an ellipse. Later Kepler extended this conclusion to other planets by analogy. Hanson agrees with Whewell that Kepler's reasoning involved the elliptic curve as a "colligating conception" in Whewell's sense, so that it was not only a "compendious expression of directly observed facts" as Mill claimed. Explanatory theories provide patterns which make data to appear intelligible, so that they constitute "conceptual Gestalts".

Hanson (*ibid*., 86) concludes that scientific discovery as inference can be described by Peirce's schema (1.18) of abduction:

(1) Some surprising phenomenon C is observed;
C would be explicable as a matter of course if H were true.
Hence, there is reason to think that H is true.

In a later paper (Hanson 1961), he argues against Popper, Reichenbach, and Braithwaite that (1) gives "reasons for suggesting H in the first place", instead of "reasons for accepting H", so that there is "a logic scientific discovery".[5]

The standard objection against Hanson is that in the schema (1) the hypothesis H occurs already in one of the premises, so that it cannot tell how H was discovered in the first place (Frankfurt 1958; Kapitan 1997). Anderson (1986), 157, notes that H is not a premise of the abductive argument (1), but it would be a premise in the explanation which corresponds to (1). Further, even though H is logically prior to

---

e.g. "The proportion .27 of the Ms are P" – is not known before the sample is drawn, so that it can be said to be "discovered by induction" (see Niiniluoto 1984, 34). Nickles (1980a) suggests that this application of the "straight rule" was the logic of discovery and the logic of justification for Reichenbach.

[5]Achinstein (1971) complains that Hanson's account does not include all relevant background information needed for discovery. The modified schema (3) below looks more promising in this respect.

the second premise "H would explain C", it is natural think that in temporal terms H and "H would explain C" are simultaneously arrived at.[6]

Another reply is to modify (1) by replacing the particular hypothesis H with a class K of hypotheses (see Hanson 1961):

(2) The surprising fact C is observed;
C would not be surprising were a hypothesis of kind K to obtain.
Hence, there is good reason for elaborating a hypothesis of kind K.

This reasoning, applied to *types* of hypotheses instead of single ones, might be an important intermediate step in the discovery of a scientific theory H which is of kind K. It has the virtue that the ultimate conclusion H is not mentioned in its premises. But a problem for (2) is that its second premise is *false* for the intermediate hypotheses of Kepler's reasoning, as they failed to agree with Tycho's observations. The following modification seems to work better than (2):

(3) The surprising fact C is observed;
there is reason to suspect that some hypothesis of kind K explains C.
Hence, there is reason to suspect that some hypothesis of kind K is true.

(See Niiniluoto 1984, 35). When the conclusion of (3) leads a scientist to think about a particular hypothesis H of kind K, and it is checked that H in fact explains C, then this schema (3) has served in the discovery of H. The second premise of (3) covers Kepler's initial but vaguely stated principles like "the sun must have something to do with causing the planets to move in their orbits" and "a general description of Mars's motion should refer to the sun as a fixed point".[7] Kepler's eighteen more specific hypotheses were of the type indicated in these principles, but the earlier ones (like circles and ovals) turned out to be unsatisfactory even though they led him closer to the final hypothesis elliptical orbit. Another example, which shows that the inference justifying the second premise of (3) may be even deductive, is Newton's famous derivation of the form of his inverse-square law of gravitation from "phenomena" (i.e., Huygens' law of circular motion and Kepler's laws).

Peirce emphasized the importance of the path along which Kepler's discovery has been reached:

> It stands upon a totally different logical footing from what it would if it had been struck out at random, or the reader knows not how, and had been found to satisfy observations. (*CP* 1.74)

This kind of approach, which follows the pattern (3), shows that proposing theoretical hypotheses at random is completely ineffective as a means to find new truths. In

---

[6]Kapitan (2014) argues that there should be a causal relation between premises and conclusion, but this is not an objection against Anderson if one approves simultaneous causality. Moreover, Kapitan leads the discussion away from discovery by formulating (1) here as an inference to the "probational" acceptance of H as the best explanation, including economic considerations of H in comparison to its competitors.

[7]Kepler's principles were motivated by metaphysical and astrological assumptions (see Burtt 1952; Popper 1963, 188).

combination with results of AI, this reconfirms Peirce's and Hanson's view that there is a logic of discovery. This does not mean that there is an algorithm for solving *all* problems: even though a universal Turing machine can in principle complete all effective calculations, given unlimited time and energy, there are recursively unsolvable problems. But AI can develop problem-solving methods for computers, together with *theories of the effectiveness of heuristic search* of a best solution within a selected domain (see Simon 1977, 326–337).[8] According to such methods, inquiry must start from "surprise", i.e. from a problem to be solved, and our first step is to restrict our attention to those hypotheses which are capable of solving our initial problem. If the problem is expressible as a why- or how-question, we should strategically limit our attention to "a problem space" consisting of hypotheses that have explanatory power with respect to the initial question. This is the basic rule of a "data-driven" logic of discovery (cf. Langley et al. 1987). In the next steps, we may look for further reasons or constraints to narrow down the space of search and then give criteria for the best solution. In retroduction, the search process is similar to geometrical analysis: a causal chain is traced by starting from its end and by reasoning backwards to earlier causes with the help of additional questions (see Sect. 2.2). Theoretical abduction can be guided by background knowledge (e.g. the mathematical form of the desired law may be inferred from earlier theories) or by simplicity considerations (e.g., the relevant laws are classified and studied by following a simplicity ordering) (cf. Langley et al. 1987; Niiniluoto 1994). A variant of such an approach to discovery is the constraint model of Nickles (1981).

N. R. Hanson was a lonely defender of the logic of discovery until his death in 1967. While researchers in AI with Herbert Simon were ready to accept a logic of discovery in the 1970s, philosophers were more hesitant (but see Achinstein 1971; Niiniluoto 1978; Anderson 1986). When "the friends of discovery" met in Reno in 1978 to discuss theories and case studies of scientific discovery, they reopened the debate on the distinction between the contexts of discovery and justification (see Nickles 1980a, b). Scientific discovery became a legitimate topic in the philosophy of science, but – especially with the influence of Larry Laudan - the dichotomy of discovery and justification was replaced by the threefold division between generation, pursuit, and acceptance.

[8]A popular theme in the AI studies in discovery is clinical diagnosis, which starts from symptoms, generates alternative medical explanations, and selects the most plausible disease as the best explanation. See Schaffner (1980), Peng and Reggia (1990), Josephson and Josephson (1994), and Magnani (2001). Another theme is forensic and legal reasoning (see Walton 2004; Gabbay and Woods 2005). Today such expert systems are capable of "deep learning" by collecting vast amounts of information from earlier studies or cases.

5.3 Abduction as Pursuit

Larry Laudan (1980), 174, argues that "both Peirce and Hanson construed the method of 'abduction' as a logic of discovery", but "it is nothing of the kind", since "it leaves that (possibly creative) process unanalyzed and tells instead when an idea is worthy of pursuit". For the notion of *pursuit*, he refers to his own study Laudan (1977), 108–114, where it means working with a research tradition which has a higher rate of progress than its rivals but which is not acceptable since it does not have the highest problem-solving effectiveness. Laudan is here appealing to his own version of the idea – modifying Thomas Kuhn's paradigm-based normal science and Imre Lakatos's scientific research programmes – that the development of science takes place in long-term evolving research traditions. For Kuhn, paradigms are not tested but rather applied to new phenomena in order to articulate them. For Lakatos, a research programme has a fixed set of core assumptions, whose change is a scientific revolution, while Laudan's traditions as successions of theories have a more flexible, historically changing core. Thus, for Laudan, the main task in pursuit is not testing hypotheses for their correctness but the application of theories in problem-solving, i.e. explanation and prediction.[9]

A different concept of pursuit was used by Martin Curd (1980), whose "logic of pursuit" belongs to "the logic of prior assessment". It does not ask, whether a hypothesis H is true or probable, but "which hypothesis should we bother taking seriously enough to test". It is concerned with a situation, where a theory has been generated but not yet developed to the point of testing (Nickles 1980a, 20). Curd does not exclude the possibility of a rational reconstruction of the logic of theory generation, but argues that Peirce's schema (1) for abductive inference should be interpreted as the logic of pursuit:

(4) The surprising fact C is observed;
The hypothesis H is capable of explaining C.
Hence, there are *prima facie* grounds for *pursuing* H.

Similarly, Kenneth Schaffner (1980) argues that Hanson's (1961) proposal that Peircean abduction gives reasons for *suggesting* a hypothesis, as distinguished from reasons for accepting a hypothesis, was ambiguous. One should distinguish between the logic of generation (i.e. the first articulation of new hypothesis) and logic of preliminary evaluation (i.e. assessment of a hypothesis for its plausibility).

---

[9]In qualitative research, an example of a research programme is the extended case method of the Manchester school of British anthropology. Tavory and Timmermans (2014) describe it as starting from one's favorite theory and extending it in the light of research data, but thereby locking theory construction into predefine conceptual boxes. The main rival of this deductive approach has been Barney Glaser's and Anselm Strauss's grounded theory, which is based on induction uncontaminated by preexisting theories and supplemented by software programs for data analysis. Tavory and Timmermans reject both of these alternatives and suggest instead "abductive analysis" which goes back and forth between surprising observations and theoretical explanations.

Unlike Laudan, Schaffner thinks that both of them are important for the logic of discovery. The former he illustrates by AI programs for clinical diagnosis.

What Curd calls "prior assessment" and Schaffner "primary evaluation" is a stage of inquiry where we try to identify which hypotheses are *plausible* or *testworthy*. The ability of a hypothesis to give at least a potentially correct answer to a question, or to give a potential explanation of the facts known so far, is an important part of the *before-trial evaluation* of a hypothesis. According to this conception, abduction is at best a route for finding worthwhile hypotheses, but does not lend any support or justification to a hypothesis. To confirm a hypothesis or to make it acceptable, we have to put it into trials or tests by deriving observable predictions from it. This stage of testing was called "induction" by Peirce in his later work after 1901.

The distinction between before-trial and after-trial criteria in the evaluation of hypotheses is in fact a classical topic in the formulations of the HD method (see Blake et al. 1960). In the mid-seventeenth century, Robert Boyle required that a good hypothesis should not assume anything absurd, be consistent, sufficient to explain at least the most important phenomena, and compatible with other known phenomena. An excellent hypothesis should also be the simplest and most explanatory, and it should predict future phenomena. Similarly, William Whewell required that a hypothetical theory ought to "explain phenomena which we have observed", but also "foretel phenomena which have not yet been observed" and are "of a kind different from those which were contemplated in the formation of our hypothesis" (Whewell 1847, 62–65). He allowed that "to try wrong guesses is, with most persons, the only way to hit upon right ones" (*ibid*., 55).

Peirce distinguished before-trial and after-trial criteria in 1903 by asking "What is good abduction?" His reply that "of course it must explain the facts" indicates that the schema (1) expresses only the minimum condition of abductive reasoning. The hypothesis must also "be capable of experimental verification" (*EP* 2:235). But Peirce discussed also additional criteria for hypotheses beyond explanatory power and testability. In 1896 he pointed out that "any scientific proposition whatever is always liable to be refuted and dropped at short notice", so that "the best hypothesis, in the sense of the one most recommending itself to the inquirer, is the one which can be most readily refuted if it is false". After this stage of falsification, the field is left free "for the main struggle" (*CP* 1.120). Peirce's most detailed account of the criteria for the choice of good hypotheses in 1901 includes again explanation of "the surprising facts we have before us" and experimental testability by new predictions, but also economic factors like cost (testability "with very little expense of any kind"),[10] caution (breaking a hypothesis up "into its smallest logical components"

[10]An interesting test case for the "economics of research" would be to study real-life decisions in science policy. Let us illustrate these ideas with current research in high energy physics. After the success of atomic theory in the first half of the twentieth century and its experimental implementation in nuclear weapons in the 1940s and nuclear power plants in the 1950s, an attempt at a deeper level of understanding was made by the quark theory of the 1960s. According to what is now known as the Standard Theory, all matter in the universe is made out of four kinds of building blocks: electrons, neutrinos, up-quarks, and down-quarks. The atomic nucleus is formed by protons and

for investigation), breath (finding one cause for different phenomena "saves repetitious work"), and incomplexity (we may start with simple hypotheses even though by complicating them they could be "brought nearer the truth") (see *EP* 2:106–114).

While early treatments of the logic of pursuit were keen on separating preliminary evaluation from generation, later commentators have been more interested in separating pursuit from acceptance. Thus, Daniel McKaughan (2008) contrasts the Generative and Justificatory interpretations with the *Pursuitworthiness* interpretation which takes Peirce's abductive reasoning "to lead to judgments about the relative pursuitworthiness of hypotheses", and such conclusions "can be thoroughly disconnected from assessments of truth-value". This agrees with the schema (4) and with Tomas Kapitan's (1997) thesis that the conclusions of abduction are practical directives of the form "it is recommended to further examine hypothesis H", based upon considerations of economy. Rune Nyrup (2015) follows McKaughan in his attempt to sidestep the "truth-connection problem" about whether better

---

neutrons, which are constituted by up-quarks and down-quarks, glued together by the strong force carried by gluons. The electromagnetic force is carried by photons, gravity by gravitons, and the weak force by W and Z particles. When the universe was born in the Big Bang, there still existed antimatter besides matter. Gravitons have not yet been discovered, but W and Z particles were found in experiments in 1982 at the European Laboratory for Particle Physics CERN.

An important addition to the Standard Theory was introduced by Peter Higgs around 1970. In order to explain why particles have mass, Higgs postulated the existence of a field and argued that mass is formed through the interaction of a particle with this field. From quantum theory and Einstein's equation $E = mc^2$ it follows that the Higgs field should be associated with (at least one) Higgs particle.

The physicists know that the Standard Theory is still incomplete, but they are vigorously pursuing its completion. Theoretical derivations are made in order to guarantee that the theory is able to account for the accepted picture of existing particles and forces. An important argument in favor of the Standard Theory is that the Higgs particle explains all other particles, while other theories fail for some known particles. In other words, the Standard Theory is worthy of further investigation, since it has more explanatory power than its rivals. To this abductive argument it is often added that this theory is also fertile in the sense that it serves as a starting point for unified field theories (such as string theory).

The pursuit of the Standard Theory also involves gigantic efforts in testing its correctness. Theoretical calculations are made in order to find testable consequences of this theory. Such tests can be made by producing beams of particles, letting them collide, detecting the results, and analyzing the data by computers. As the production of heavy particles requires high energy levels, the design of relevant experiments and the constructions of needed equipment require advanced and expensive technology. To settle the most important open problem about the existence of the Higgs particle, the CERN Laboratory replaced its Large Electron Positron collider (LEP), which resides in a circular tunnel of the length of 27 km, with a new accelerator, the Large Hadron Collider (LHC). Its activity started in 2007 with more than 4000 scientists collaborating in the hunt for the Higgs particle. After running the experiment with huge amounts of data, the physicists announced on July 4, 2012, the existence of the Higgs boson. The argument can be understood as a complex abduction, where the collected data constitutes the "surprising facts", and the existence of Higgs boson is its best explanation. In this way, the Standard Theory was confirmed and found acceptable by the physics community; otherwise, alternative theories about the constitution of matter should have been investigated. Even though the Standard Theory does not have immediate practical applications, this value of the information from the CERN laboratory was to enough to compensate the high economic investments and costs.

explanations are more likely to be true. He develops a formal decision-theoretic model to show that, "all else being equal", if H_1 is more explanatory than H_2, we are justified in pursuing H_1 rather H_2. For Nyrup, pursuit of a hypothesis H means an attempt to find out the truth value of H, but his argument does not presuppose any connection between the explanatoriness and the truth of H.[11]

Even though Dov Gabbay and John Woods (2005) do not use the term "pursuit", their GW model of abduction belongs to this camp (cf. Sect. 3.4). Their starting point is an agent's "ignorance problem" about "whether this-or-that is the case or whether such-and-so is the right decision" (see Woods 2007). This problem is not solved by the agent's present knowledge K, nor by a hypothesis H alone, but were H the case then the revision of K by H would attain the agent's cognitive target. Such "presumptive" attainment of a cognitive target leaves the ignorance-problem unsolved: abduction does not secure us knowledge, and "does not warrant belief" at all, so that abduction is *ignorance-preserving*. Thus, the GW model is not interested in the original discovery of a hypothesis H, and explicitly denies the "evidential clinching" power of abduction to justify H. Still, at least when H satisfies further conditions of relevance and plausibility, H is pursued in the sense that it is "put to function": in "partial" abduction H is presented as a conjecture, and in "full" abduction H is used as a basis of new action. For example, in courts of law, it is typical that "jurors find an accused guilty of the offence he has been charged", but "they do not know whether the offence was committed by him" (*ibid*,. 316).

Full abduction in the GW model interprets pursuit of a hypothesis as the decision to use H in action. This is an important topic, not developed by Peirce in any detail.[12] But the GW suggestion that it could be rational to "put a hypothesis H to function" in action in complete ignorance, i.e. without having any degree of belief in H,[13] is in flat contradiction with Peirce's insistence that his pragmatism is "scarce more than a corollary" to Alexander Bain's definition of belief as "that upon which a man is prepared to act" (*CP* 5.12). A rational account of human conduct should take into account the beliefs and values (risks) of the agent. A sophisticated approach is given

[11]Nyrup's argument is a variant of cognitive decision theory (see Levi 1967). Apart from the question, whether his equality assumptions are realistic, his argument (if valid) would be applicable to any epistemic value, so that it fails to illuminate Peirce's abductive or explanatory reasoning.

[12]Peirce mentions the "practical" aspect of abduction as affecting voluntary conduct in his Manuscript 318 in 1907. In Manuscript 637 in 1909 he states that abductions must "recommend a course of action". See Kapitan (1997), 493; *CP* 8.322. This emphasis might have been a basis for a "Deweyan abduction", given John Dewey's attempt to reduce logic to a theory of problem-solving (see Paavola 2015).

[13]Barés Gómez and Fontaine (2017) try to illustrate the GW model of action by an example, where two friends in Mexico City observe that their workmate succeeds in reaching the university before them by disembarking the train in a station from which no direct connection to the university is available by their underground map. In the next day they do not take the longer route but follow their guess that such a connection is just missing from their map. Is this an instance of action in persisting ignorance? On the contrary, the workmate's surprising success clearly increases the probability of the how possible -explanation that the map is incomplete and another connection is available (cf. Sect. 6.2).

by Bayesian decision theory, which recommends the maximization of expected utility, but the most classical treatment is Aristotle's practical syllogism, which concludes an action on the basis of the agent's wants (intentions) and beliefs about the necessary conditions of the ends. Risto Hilpinen (2007) writes this inference in a form of "practical abduction" whose structure matches with Peirce's schema (1) and its variant (4) for pursuit:

(5) I wish to attain the end E.
If I were to perform action A, E would be the case.
Hence, there is reason to perform A.

A similar proposal for "projectual abduction" is given by Tuzet (2006). In this kind instrumental reasoning, the starting point is not a surprising fact but rather a goal, and the abductive conjecture expresses means to realize this goal (cf. Schurz 2017, 170).[14] If Magnani (2009) wishes to include practical reasoning as an important element of his "manipulative abduction" (cf. Park 2017a), the schema (5) would give a promising starting point.

In spite of the arguments discussed above, it is not quite clear that pursuit and acceptance can always be sharply separated from each other. Evidence for this can be found in Peirce's own statements. In the midst of his "economic" criteria for good hypotheses we find the following:

> When we are in doubt which of two hypotheses ought to have precedence, we should try whether, by enlarging the field of facts which they are to explain, a good reason will not appear for giving one of them a decided preference over the other (*EP* 2:114).

This advice of finding more evidence is very appropriate for inferences in legal courts, forensic science, and the determination of causes of death in pathology.[15] Cleland (2002) illustrates the same idea by showing how additional evidence served as a "smoking gun" in the assessment of rival historical hypotheses about continental drift and the extinction of the dinosaurs. The search for additional evidence may involve new instruments for measurement and experimentation (such as Galileo's

---

[14]This kind of pragmatic abduction is important also in futures studies, which builds anticipatory scenarios or strategic action plans in order to reach desirable goals or to avoid undesirable threats (see Niiniluoto 2009). The reverse form of forecasting, or moving backward in time from a desirable future to present policy-measures which will help to attain the specified future, is called "backcasting". This term was proposed in 1982 by John Robinson in the context of energy policy.

[15]The advice of expanding evidence is also relevant to the problem of empirical underdetermination of theoretical hypotheses in science (see Niiniluoto 1999a, 176–177). This antirealist argument is usually presented in the form that for any empirical evidence E and for each theory T there is another different theory T′ such that they are empirically equivalent with respect to E (see e.g. Psillos 1999, 164). A stronger formulation allows E to include all empirical truths, so that theories T and T′ have the same empirical content, i.e. T and T have the same deductive relations to the observational language L_0. But the situation changes dramatically, if probabilistic relations to L_0 are also taken into account. Moreover, language L_0 is not absolute and fixed, but can be expanded by using new instruments. More generally, theory choice in science takes place in the context of expanding empirical evidence, together with theoretical background assumptions, including "bridge principles" between theoretical and observational terms.

telescope and van Leeuwenhoek's microscope). Peirce himself warned that one should never maintain that something can never be known: when Comte claimed that the chemical composition of fixed stars cannot be attained, "the ink was scarcely dry upon the printed page before the spectroscope was discovered" (*CP* 1.135).

It is natural to interpret this advice of *expanding evidence* as an effective guide in truth-seeking. It also points out the fact that making observations and experiments at random is ineffective as an inductive method, at least if we do not know what we are seeking.[16] Peirce was familiar with the idea that good experiments are "questions put to nature" (see *EP* 2:215), i.e. they have to be designed so that they give answers to our requests of important information.[17] And observational facts are surprising only relative to tentatively accepted background theories.

In his 1910 reflections on "The Doctrine of Chances", Peirce defined the *plausibility* of a theory in the following way:

> By Plausible, I mean that a theory that has not yet been subjected to any test, although more or less surprising phenomena have occurred which it would explain if it were true, is in itself of such a character as to recommend it for further examination or, if it be *highly* plausible, justify us in seriously inclining toward belief in it, as long as the phenomena be inexplicable otherwise (*CP* 2.662).

McKaughan (2006), 454, quotes the first part which support his case for pursuitworthiness, but omits the last part which is related to justification. Elsewhere in 1908 Peirce stated that holding a hypothesis to be plausible ranges "from a mere expression of it in the interrogative mood, as a question meriting attention and reply, up through all appraisals of Plausibility, to uncontrollable inclination to believe". The "first stage of inquiry", with a characteristic formula of Retroduction, i.e. reasoning from consequent to antecedent, covers "the whole series of mental performances between the notice of the wonderful phenomenon and the acceptance of the hypothesis" (*CP* 6.469; *EP* 2:441).

In the next chapter, we shall reconsider the distinction between pursuit and acceptance, or before-trial and after-trial evaluation of theoretical hypotheses, within the probabilistic Bayesian framework of confirmation relations. We shall argue that abduction is not ignorance-preserving but rather credence-increasing inference.

---

[16]This is a valid point in Hempel's (1966) critique of inductivism.

[17]See the summary of Hintikka's interrogative model of inquiry in Sect. 3.5.

Chapter 6
Abduction and Confirmation

Abstract In this chapter we go beyond Peirce by studying the role of abduction in the weak justification or confirmation of hypotheses. Section 6.1 deals with qualitative confirmation, or its special cases of inductive and abductive confirmation as identified by Howard Smokler. Section 6.2 treats quantitative abductive confirmation within the setting of Bayesian epistemic probabilities. The basic theorem of abductive confirmation shows that contingent evidence E is positively relevant (PR) to consistent hypothesis H if H logically entails E. More generally, if H deductively or inductively explains E, then E PR-confirms H. This result shows that abduction is generally credence-increasing, and refutes the GW-thesis that abduction is ignorance-preserving. In Sect. 6.3 we show comparatively that better explanations with higher explanatory power receive stronger degrees of confirmation. Thus, it is useful to follow Carl G. Hempel in combining the notions of explanatory and predictive power into one concept of the "systematic power" of a scientific theory. Section 6.4 analyzes the virtues of unification by theories. It is shown that Whewell's "consilience of inductions" (i.e. a theory entails two independent laws) has to be distinguished from the idea of inductive systematization (i.e. two independent empirical phenomena become relevant to each other relative to a theory). The latter case is covered by Myrvold's measure of unification.

In this chapter we go beyond the pursuit interpretation by studying the role of abduction in the weak justification or confirmation of hypotheses. Section 6.1 deals with qualitative confirmation, or its special case of "abductive confirmation" as identified by Howard Smokler. Section 6.2 goes beyond Peirce by treating quantitative abductive confirmation within the setting of Bayesian probabilities. We prove a basic theorem which states that explanatory success is probability-increasing. In investigating the relation between explanation and confirmation, we shall see that it is useful to follow Carl G. Hempel in combining the notions of explanatory and predictive power into one concept of the "systematic power" of a

I. Niiniluoto, *Truth-Seeking by Abduction*, Synthese Library 400,
https://doi.org/10.1007/978-3-319-99157-3_6

scientific theory (Sect. 6.3). Section 6.4 analyzes the virtues of consilience or unification by theories.[1]

6.1 Inductive and Abductive Confirmation

Practically speaking all philosophers of science agree that scientific reasoning has an *empirical basis* which can be expressed by observational statements. But a lively debate on the nature of this empirical basis took place in the Vienna Circle in the 1930s. For the *phenomenalists*, perceptual statements are reports of immediate experiences (e.g. "Something red and green appears to me now"), while for the *physicalists* they are about ordinary objects and facts in our environment (e.g. "I see an apple in front of me"). Physicalist statements can be obtained by observation and measurement, possibly aided by instruments, and by experiments, where the researcher manipulates nature.[2] Singular statements about particular individuals or events may be verifiable or falsifiable by observations. But following Peirce, there is an abductive element in such statements (cf. Sect. 4.1), so that his fallibilism extends to the empirical basis as well. The physicalist wing of the logical empiricists (Otto Neurath, later Carnap) agreed that the statements in the observational language are more or less uncertain.[3]

Classical positivists (Auguste Comte, Ernst Mach) in the nineteenth century and some logical positivists in the 1920s (Rudolf Carnap) restricted the task of science to the level of empirical generalizations. But most of the interesting scientific hypotheses transcend the empirical basis in several ways. Universal statements can be falsified by empirical counterexamples, as Karl Popper emphasized in 1934 (see Popper 1959), but in unlimited domains they cannot be verified by any finite number of instances. This is the *problem of induction*. Statements about theoretical entities behind the observational phenomena, including statements about past historical events, can be tested by observing their deductive and probable consequences. This is the *problem of abduction*. Peirce's account of ampliative inferences was intended as a solution to both of these problems. Induction proceeds from empirical

[1]This Chapter is partly based on Niiniluoto (2004, 2016).

[2]Bas van Fraassen's (1989) "constructive empiricism" restricts the empirical basis to observations by our "naked" senses.

[3]In all sciences, statements already in the empirical level are conceptually laden interpretations of reality. In the human and social sciences, the situation is even more complex, since the empirical basis typically concerns human agents and their actions which are meaningful to the agents themselves. Thus, the researcher has to impose a second-order meaning to the first-order discourse of the agents. For an application of Peirce's semiotics to this problem of meaning construction, see Tavory and Timmermans (2014).

data to generalizations, abduction from effects to causes or from surprising facts to their theoretical explanation (see Chap. 1).[4]

The problem of induction was pressing for all empiricists, and intensively discussed in the first half the twentieth century. It was realized that inductive generalization cannot be justified by reducing it to deduction, where the premises logically entail the conclusion. So the notion of logical entailment was replaced with a new relation of *confirmation*, *inductive support*, or *factual support* between the empirical premises of induction and its conclusion. Nicod's criterion, which stated that a generalization of the form "All Fs are G" can be confirmed only by its positive instances "Fa&Ga", led to difficulties with Hempel's paradox of the ravens. In the 1940s authors like Janina Hosiasson-Lindenbaum and Carnap (fiercely opposed by Popper) sought Bayesian solutions by introducing quantitative degrees of confirmation which use probability theory as its tool. This led to the formal systems of inductive logic by Carnap and Jaakko Hintikka (see Niiniluoto 2011c) and several other attempts to define such degrees (see Foster and Martin 1966).

The problem of abduction presupposes that the scientific claims about the unobservable are genuine statements with a truth value. This realist view was denied by the *instrumentalists*, like Pierre Duhem and Henri Poincaré, who take theories to be just more or less convenient conceptual tools for summarizing observations or "saving the appearances".[5] This is one reason why abduction became an important theme only in the 1950s when philosophers of science (like Herbert Feigl and Wilfrid Sellars) started to defend the realist interpretation of scientific theories (cf. Niiniluoto 1999a).

Already the 1935 International Congress of Unified Science in Paris was an important watershed. Alfred Tarski announced his semantic definition of absolute truth, and convinced Popper about the tenability of the objective correspondence theory of truth (see Popper 1963). Carnap presented a paper, which made a clear difference between timeless truth and the time-dependent and evidence-relative notion of "confirmation".[6] In his article "Testability and Meaning" in 1936, Carnap took an important step away from logical positivism by replacing the aim of verification by confirmation and by allowing theoretical statements which are not reducible to the observational language by explicit definitions.

---

[4]Lawlike statements, including probabilistic laws, make assertions about counterfactual situations. The doctrine that there are real possibilities beyond the domain of actuality was important for Peirce′'s later realist versions of his pragmatism (see Peirce 1901). This inferential problem has to combine the method of abduction with experimental operations which also realize counterfactual possibilities.

[5]This idea goes back to Plato's Academy and Greek astronomy, and was hotly debated during the time of the Copernican Revolution. An excellent history, with sympathies on the side of the instrumentalists, is given by Duhem (1969).

[6]Carnap called this concept "Bewährung" following Popper, but Popper's "Bewährungsgrad" was later translated as "the degree of corroboration" (see Popper 1959), to distinguish it from the inductivist notion of "confirmation".

Carl G. Hempel, who had shared Neurath's commitment to the coherence theory of truth, also adopted the Tarskian notion of truth. In his studies in the 1940s, he demanded in the spirit of realism that satisfactory scientific explanations must be true. Only in the notion of *potential explanation* we may have explanations with unknown truth values. Thus, the aim of explanation is not only to save the appearances, but to subsume them under true regularities in the world. This is in harmony with the old Aristotelian "regressive" idea that science wishes to proceed from *knowing that* to *knowing why*: once we know that grass is green, we would like to have a deeper insight into the causes of this fact, i.e. we wish to know why grass is green. This desire is fulfilled by an explanation of the fact that grass is green (cf. Sect. 2.1). A similar distinction was made by Hempel (1965) between *reason-seeking* why-questions (what reasons do we have for thinking or claiming that C?) and *explanation-seeking* why-questions (why is it the case that C?). In the logic of questions, the statement that C is taken to be a presupposition for the question "why C?".

Wesley Salmon (2001a, 79) argued that the characterization of non-demonstrative inference as abduction or inference to the best explanation serves "to muddy the waters" by fostering confusion between Hempel's two types of why-questions. However, if we infer H on the basis of its ability to explain a known fact C, we are indeed explaining why C in Hempel's sense, but we are not thereby proposing new reasons for believing that C, but rather seeking abductive reasons for believing that H. Similarly, reasons for believing that H should be distinguished from the further question "Why H?". Hence, no confusion between Hempel's two questions is involved here.

Hempel (1966) combined his approach to explanation with the hypothetico-deductive method of scientific inference, so that he did not open the discussion on Peircean abduction. But Hempel was ready to use the concept of confirmation in connection with the HD inference: if the deductive test predictions E from a hypothesis H turn out to be true, then H is confirmed by the evidence E, and otherwise E refutes or disconfirms H.[7] This is an extension of the notion of confirmation from inductive generalization to HD testing – or, more generally, to a relation between a hypothesis and its evidence.

Hempel's important contribution in the 1940s was the idea of analysing the concept of empirical confirmation by qualitative or structural principles (see Hempel 1965; cf. Sect. 3.2). One intuitive notion is to regard confirmation as a weakening of the relation of logical consequence or entailment: according to the *Entailment* condition, if an observation report entails a hypothesis, it also confirms the hypothesis.

(E) If evidence E logically entails hypothesis H, then E confirms H.

[7]Note that the notion of confirmation or support is weaker than the notion of *acceptance*, where a hypothesis is so strongly supported by evidence that it is rational to tentatively accept it as true, i.e. to include it in the body of what is currently taken to be scientific knowledge. See Chap. 7 for the stronger notion of justification by IBE.

As entailment itself is transitive, one may also suggest that confirmation is transmitted to logical consequences. This is the *Special Consequence* principle:

(SC) If evidence E confirms hypothesis H, and K is entailed by H, then E confirms K.

Another idea is related to confirmation by HD testing, where we check whether the observable consequences of a hypothesis are true or not. Thus, according to the *Converse Entailment* condition,

(CE) If hypothesis H logically entails non-tautological evidence E, then E confirms H.

For example, if theory H achieves *deductive systematization* between empirical statements E and E′, i.e. $H \,\&\, E \vdash E'$ and not $E \vdash E'$, then $H \vdash (E \rightarrow E')$, so that $E \rightarrow E'$ confirms H by CE (see Hempel 1965). By the *Converse Consequence* condition, evidence confirming a hypothesis also confirms all logically stronger hypotheses:

(CC) If K logically entails H and E confirms H, then E confirms K.

Hempel made the important observation that CC and SC are incompatible, since a notion of confirmation satisfying both of them would be trivial in the sense that any statement confirms any other statement. The same holds for the pair CE and SC (see Hesse 1974).

The incompatibility of conditions of CE and SC has an important consequence to the once popular transitivity interpretation of Whewell's consilience of inductions (see Hesse 1974, 141–146; Niiniluoto and Tuomela 1973, 228–230).[8] Assuming that a theory H entails empirical statements or regularities E and E′, some authors suggested that direct evidence for E counts via H as indirect evidence for E′ (see Kneale 1949, 108; Nagel 1961, 64–65.). However, such an argument would apply first CE in concluding that E confirms H and then SC in concluding that E confirms E′.

Howard Smokler (1968) noted that E and SC are typical in enumerative and eliminative induction, where a generalization receives empirical support from its positive instances. For example, if all ravens observed so far support the generalization that all ravens are black, they also support the consequence that the next raven is black. On the other hand, he proposed that CE and CC are satisfied by "abductive inference", where (following Peirce) a hypothetical theory is supported by its power to explain surprising phenomena. Thus, Hempel's proof that these abductive and enumerative principles of confirmation cannot reasonably be satisfied together gives further motivation to Peirce's original thesis that abduction and inductive generalization are two different kinds of ampliative inference (cf. Sect. 1.5).

[8]We shall see below that this was not Whewell's own interpretation, as consilience for him provided an argument in favour of the unifying theory H.

The straightforward definition of *deductive confirmation* (E confirms H iff H entails E) would satisfy Smokler's conditions CE and CC.[9] In the spirit of the HD method, its special case is *predictive confirmation*:

(PC) E predictively confirms H iff H deductively predicts E.

But these adequacy conditions CE and CC have been criticized for allowing confirmation too easily, since entailment is monotonic and thus admits arbitrary strengthening of the premise: if H $\vdash$ E, then H&K$\vdash$ E for any K; hence, if E deductively confirms H, it also deductively confirms H&K for any K. More plausible versions of these principles – call them CE* and CC* – are obtained by replacing deductive entailment $\vdash$ by the stronger condition of deductive explanation (see Sect. 3.2). For example, CE* can exclude some cases where the entailment is too trivial (self-explanations) or *ad hoc*.[10] Define *abductive confirmation* by

(AC) E abductively confirms H iff H deductively explains E,

where H is consistent and E is non-tautologous (see Niiniluoto and Tuomela 1973, 227). This can be suggested as an interpretation of Peirce's canonical schema (1.18) for abduction, where the explanation of a surprising fact E by a hypothesis H gives "reason to think that H is true". Then confirmation defined by AC satisfies CE*, but not generally CC* without additional conditions (see the monotonicity principle (3.11′)). While AC is in line with Peirce's account of abduction, it should be generalized to cover cases of inductive-probabilistic explanation as well.

If Smokler's notion of abductive confirmation is defined by the structural conditions CE and CE, then this wide conception is unable to distinguish the cases of explanation (AC) and prediction (PC). The narrower conception of abductive confirmation (cf. AC) would include a link to explanation. But then one should ask, whether the idea of confirmation is really applicable to explanatory success. Namely, some philosophers who claim that abduction has no role in the context of justification – as opposed to discovery and pursuit (see Chap. 5) – have emphasized the distinction between explanation and prediction, so that they are willing to include only predictive success in the context of confirmation. In other words, they accept PC but reject AC. Here they can appeal to Peirce's later methodological view which takes abduction to be "the first step of scientific reasoning", only later followed by "inductive" testing (see Sect. 1.5). Along these lines, Salmon challenges the "explanationist" view that explanatory virtues are relevant to confirmation: "confirmation is logically independent of explanation" (Salmon 2001a, 88). Bas van Fraassen (1989) makes a distinction between "confirmational" and "informational

[9]The concept of conditional deductive confirmation of H by E can be defined by the requirement that there is an observational statement C such that H achieves deductive systematization between C and E, i.e. H&C$\vdash$ E. To avoid the trivial choice of C as H $\rightarrow$ E, one should require that C is logically independent of H (see Kuipers 2000, 36).

[10]Laudan (1990) also argues that hypotheses are not confirmed by all of their observational consequences.

virtues" of theories, arguing that explanatory power belongs to the latter type. In the same way, it is sometimes claimed that *accommodation* (i.e. the adoption of an explanatory hypothesis to account for already known data) has to be sharply distinguished from *prediction* in the sense that accommodation does not have any power of confirmation.[11] We shall return to this issue in the next section. Here one may note against the critics of explanationism that some historically important theories (Copernicus, Darwin) have received their strongest support from their ability to explain "old evidence", which was not predicted by the theory but known already before its proposal. For some currently accepted theories, e.g. dark matter, the only evidence comes from their explanatory power. Moreover, we should not overlook the possibility – to be examined in the Bayesian framework in Sect. 6.3 – that the overall confirmation of a theory may be a combination of its explanatory and predictive power.

6.2 Bayesian Confirmation

In his early work, Peirce treated induction and hypothesis as forms of *probabilistic* reasoning by associating a mode of argument with is its *truth-frequency*, i.e. its ability to yield true conclusions from true premises. Peirce's program of probabilistic inference is realized in statistics by the error probabilities of the Neyman-Pearson tests of significance.[12]

Even though Peirce as a frequentist was sharply critical of the classical *Bayesian* theory of probabilistic inference (see e.g. *EP* 2:215), and thought that induction does not lend probability to its conclusion (*CP* 2.780), we shall see in this Section that Bayesianism provides a useful framework for studying abduction and induction as forms of ampliative reasoning.[13] In this school, Bayes's Theorem has traditionally been used for the calculation of "inverse probabilities", i.e. the probabilities of causes given effects. Bayesian confirmation theory (Stanley Jevons, J. M. Keynes) was conceived as an account of induction as converse deduction. These ideas are clearly linked with Peirce's characterizations of hypothesis as inference from effects to causes.

Assume that a probability measure P is available for the relevant language. For example, E may be (a statement describing) the event of a person's death, and H_i

[11]See Howson and Urbach (1989) and Lipton (2004).

[12]More precisely, in a simple NP-test with null hypothesis H_o and a specific alternative H_1, the significance level α is the statistical probability of rejecting H_o when H_o is true, so that $1 - \alpha$ is the truth-frequency for inferences about H_o. The power of the test is the statistical probability of rejecting H_0 (or accepting H_1) when H_1 is true, so that it is the truth-frequency for inferences about H_1. This note corrects a mistake in Niiniluoto (1999b), 447.

[13]Psillos (2004), who wishes to justify abduction as IBE in a non-probabilistic way, is skeptical about such an approach (cf. discussion by Iranzo 2007). Lipton (2004), in the second edition of his book, extended his account to promote "friendship" between Bayesianism and IBE.

($i = 1, \ldots, n$) describe the possible causes of her death. Given statistical mortality tables, let the prior probability $p_i = P(H_i)$ be the relative frequency of the occurrence of event H_i, and let the likelihood $q_i = P(E/H_i)$ be the probability that H_i produces the effect E. Then *Bayes's Theorem* gives the posterior probability of the cause H_i given E:

(1) $P(H_i/E) = p_i\, q_i/P(E)$,

where

(2) $P(E) = \sum_{j=0}^{n} p_j\, q_j$.

Such posterior probabilities (1) are "abductive probabilities" in a Peircean sense, since they proceed from effects to causes, as it were.

The same formula (1) can be used when H_i is a hypothetical theory. In this case, the probabilities of statements H_i and E are usually interpreted as *epistemic probabilities*: $P(H_i)$ is our initial *degree of belief* in the truth of H_i, and $P(H_i /E)$ is this degree of belief or *credence* of H_i given evidence E. Inferences are thus reconstrued in terms of epistemic probabilities which express coherent degrees of belief of rational agents. Bayes's Theorem tells how prior probabilities P(H) of hypotheses are transformed to posterior probabilities P(H/E) via the likelihood P(E/H). While for subjectivists such probabilities are personal degrees of belief, systems of inductive logic include more objective ways of allocating probabilities to statements expressible within a language (for an overview, see Niiniluoto 2011c). The Bayesians can also employ objective physical probabilities especially in their treatment of likelihoods P(E/H) and probabilistic laws. Note that the application of Bayes's Theorem presupposes that the hypotheses H_i constitute an "ultimate partition" (Levi 1967) of mutually exclusive and jointly exhaustive statements. When alternative hypotheses H_i are not exhaustive, the ultimate partition is achieved by adding the "catch-all hypothesis" $\sim(H_1 \vee \ldots \vee H_n)$. The simplest partition with two alternatives is of the form {H, ~H}.

Inspired by Keynes's logical interpretation, Carnap in 1950 called the posterior probability P(H/E) of a hypothesis H given evidence E the "degree of partial entailment" or the "degree of confirmation" of H by E. After Popper's (1959) criticism, in the Second edition of his work Carnap (1962) made a difference between confirmation as posterior probability and confirmation as increase of probability. This gives two probabilistic definitions of confirmation:

(HP) *High Probability Criterion*: E HP-confirms H iff P(H/E) is sufficiently large.

This means that P(H/E) should be at least as large as a fixed threshold value $q \geq ½$.

(PR) *Positive Relevance Criterion*: E PR-confirms H iff $P(H/E) > P(H)$.

PR is equivalent to conditions $P(E/H) > P(E)$ and $P(H/E) > P(H/{\sim}E)$. Further, E *disconfirms* H if $P(H/E) < P(H)$, while E is *irrelevant* to H if $P(H/E) = P(H)$. Both HP and PR can be relativized to some background knowledge B.

It immediately follows that what Smokler called "enumerative" principles E and SC are satisfied by the High Probability criterion – also by Hempel's own "satisfaction criterion". If inductive-probabilistic (IP) explanation is defined by the condition of high probability of the explanandum given the explanans (see Hempel 1965), then we have

(3) If H IP-explains E, then H HP-confirms E but not conversely.

On the other hand, Smokler's "abductive" principles CE and CE* are satisfied if confirmation is defined by the Positive Relevance criterion. However, positive relevance does not generally satisfy the controversial converse consequence condition CC. According to Bayes's Theorem,

(4) $P(H/E) = \frac{P(H)P(E/H)}{P(E)}$.

Now if H entails E, we have P(E/H) = 1. Hence, by (4),

(5) If H logically entails E, and if P(H) > 0 and P(E) < 1, then P(H/E) > P(H).

This result is completely general in the sense that it is valid for all epistemic probability measures P and for all non-zero prior probabilities P(H) > 0. It allows H to be a strong and informative theory with theoretical concepts. Hence, CE* is satisfied as well:

(CE*) If H and E are contingent statements, and H deductively explains E, then E PR-confirms H.

More generally, as positive relevance PR is a symmetrical relation, it is sufficient for the confirmation of H by E that H is positively relevant to E. For example, if an infection H increases the probability of fever E, then the fever supports the hypothesis of infection. If inductive explanation is defined by the positive relevance condition, i.e. H inductively explains E if and only if P(E/H) > P(E),[14] then

(CE**) If H is a positively relevant inductive explanation of E, then E PR-confirms H.

Combining these results, we have

(6) If H deductively or inductively explains E, then E PR-confirms H.

Thus, the Bayesian approach immediately justifies the idea that *explanatory success is confirmatory or credence-increasing*. This a generalization and justification of the principle AC of abductive confirmation. But by the same arguments we can establish a generalization of the principle PC of predictive confirmation:

(7) If H deductively or PR-inductively predicts E, then E PR-confirms H.

A more complex situation obtains when the test statement E is uncertain, so that it cannot be used as evidence in conditionalization P(H/E). Suppose, however, that J is

[14]This is the basic idea Salmon's SR-model. See Niiniluoto and Tuomela (1973), Festa (1999).

a reliable indicator of E in the sense that there are constant probabilities P(J/E) = p and P(J/¬E) = q, where p > q (see Bovens and Hartmann 2003). The reliability condition p > q is equivalent to P(J/E) > P(J), so that J is positively relevant to E. As positive relevance does not generally satisfy CC, we cannot conclude from this condition and H⊢ E that J is positively relevant to H as well. But this inference is warranted, if it can be further assumed that H is irrelevant to J. Hence, as a modification of (5) for uncertain evidence, we can prove the following: if H logically entails E, P(H) > 0, J is a reliable indicator of E, and P(J/H&E) = P(J/E), then P (H/J) > P(H).

The results (5) and (6) express *the Basic Theorem of Abductive Confirmation* (see Niiniluoto 1999b). This simple application of Bayes's Theorem is sufficient to show that Peirce's schema of abductive reasoning is truth-conducive in the positive but weak sense that explanatory success increases the probability that the explanatory theory is true. This gives an answer to the characterization problem: while deduction is truth-preserving, and the suggestion that abduction as ignorance-preserving is found implausible (cf. Gabbay and Woods 2005), now we can see that the common logical feature of abductive arguments is that they are *probability-increasing* or *credence-increasing*. By (7), they share this feature with successful prediction. The strength of the quantitative Bayesian treatment is thus its ability to account for the most central qualitative principle CE (and CE*) of abductive confirmation.[15] A positive solution to what Nyrup (2015) calls the "truth-connection problem" is thereby obtained (see Sect. 5.3).[16] Further, we have found reason to reject those

[15]Gerhard Schurz (2008a) defends abduction as a mode of inference with some value in justification, but rejects the Bayesian approach as being unable to demarcate scientifically worthwhile hypotheses from pure speculations. To show that Bayesian incremental confirmation is too easy, Schurz (2008b) considers the God's will hypothesis

(G-E) God wants E, and whatever God wants, happens.

where E is any empirical phenomenon. As G-E deductively entails E, by (5) E confirms G-E, even though G-E is purely speculative and should not receive any scientific confirmation. This argument raises many issues that belong to the philosophy of religion. All theologians would not accept that evil events happen by God's wishes. Some religious thinkers, like Richard Swinburne, have applied Bayesian confirmation theory to the hypothesis that God exists. So the Bayesian approach is flexible enough to reconstruct the thinking of a religious person who sees divine providence everywhere and thereby finds confirmation for his or her faith. One may recall that confirmation is a weak epistemic concept which does not yet guarantee that a hypothesis is acceptable as true (Niiniluoto 2008). But I agree with Schurz that G-E should not receive any scientific confirmation. However, this does not require that Bayesianism is rejected, as Schurz pleads, since a Bayesian can block the confirmation of such speculative religious hypothesis by giving them zero prior probability. This move could be justified by arguing that problems with the notion of God's omnipotence show that G-E is inconsistent. Another line of argument would be to accept that G-E entails E, but deny that G-E explains E, since it includes an *ad hoc* assumption that God wants E, so that G-E is not supported by the refined result (6).

[16]The Basic Theorem gives a Bayesian counterargument to the thesis of Roche and Sober (2013) that "explanatoriness is evidentially irrelevant". They defend this thesis by claiming that P (H/E&B) = P(H/E), when B states "were H and E true, H would explain E". But our theorem

criticisms of explanationism which deny that explanatory success is relevant to confirmation. Against van Fraassen's (1989) thesis that explanatory power is only an "informational" virtue, result (6) shows that abduction is not only governed by the search of new information: abduction seeks theories with explanatory information, and such theories receive some probabilistic confirmation from the explained data.

The Basic Theorem gives an argument against some sceptical anti-realists who have suggested that the success of scientific theories in explanation and prediction does not in any way indicate that these theories are *true* (in the realist sense of correspondence with reality) or that the postulated theoretical entities exists in reality. For example, Bas van Fraassen (1980) proposes to conclude only that successful theories are *empirically adequate* (i.e. all of their observable consequences are true). Larry Laudan (1990), who thinks that truth is a utopian aim of science, is still willing to speak about the empirical support or confirmation of theories, but for him such support of a theory concerns only its *reliability* (i.e. true expectations about its next observable predictions).

The Basic Theorem can be generalized to interesting comparative versions (see Kuipers 2000). Peirce knew that from a testworthy hypothesis one should deduce "whatever experiental predictions are extremist and most unlikely" (*CP* 7.182). The Bayesian approach shows that from such severe tests, where P(E/B) is low, we can gain more PR-confirmation or corroboration to our hypothesis (cf. Popper 1963, 391).

But if the probability P(E/B) is one, then P(H/E&B) = P(H/B) and no learning from such evidence E is possible in Bayesian terms. Still, genuine confirmation should be possible at least if it was not known that H explains E. For example, the great achievement of Copernicus was to show what the heliocentric hypothesis really can explain. The simplest solution to this "problem of old evidence" is to reformulate the situation so that the probability P(E/B) is not one. Alternatively one can drop the idealized assumption that $H \vdash E$ implies P(E/H) = 1.[17]

6.3 Explanatory Power and Confirmation

A natural and useful way of refining the Basic Theorem is to introduce quantitative degrees of explanatory power and degrees of confirmation, with the aim of showing comparatively that *better explanations receive stronger confirmation* (see Niiniluoto 1999b):

(6) is expressed as a conditional "if B then P(H/E) > P(H)". If B is a proposition which can occur as argument of the probability function, then the correct counterpart of (6) is P(H/E&B) > P(H/B), i.e. E is positively relevant to H given B, rather than the equation of Roche and Sober. For a different reply, see Climenhaga (2017).

[17] For treatments of this problem, see Howson and Urbach (1989), 275–284, Niiniluoto (1983), and Earman (1992).

(8) If K is a better explanation of E than H, then E confirms K more than H.

To prove this result, we need a quantitative definition of the *explanatory power* expl (H,E) of H with respect to E, and of the *degree of confirmation* conf(H,E) of H by E.[18] Popper (1959) attempted to achieve this directly by defining his measure of corroboration as a function of explanatory power.

The first measure of *explanatory power* of theory H with respect to evidence E was defined by Hempel in 1948 as

$$\mathrm{expl}_1(\mathrm{H,E}) = \mathrm{P}(\sim \mathrm{H}/ \sim \mathrm{E})$$

(see Hempel 1965). Here E may be a conjunction of several empirical statements. Hempel derived this measure as the ratio between the common information content of H and E (i.e. 1- P(H v E)) and the content of E (i.e. 1 - P(E)). Another definition, due to Jaakko Hintikka (1968), is

$$\mathrm{expl}_2(\mathrm{H,E}) = \frac{P(E/H) - P(E)}{1 - P(E)}.$$

A related measure was proposed by Håkan Törnebohm in 1966:

$$\mathrm{expl}_3(\mathrm{H,E}) = \frac{logP(E) - logP(E/H)}{logP(E)}.$$

All of these measures receive the maximal value one if H deductively explains (the whole of) E, so that they can primarily be used for the comparison of non-deductive explanations. According to expl_2 and expl_3, K is a better explanation of E than H if and only if the likelihood of K is larger than that of H, i.e. P(E/K) > P(E/H). According to expl_1, K is a better explanation of E than H if and only if P(K)(1 – P (E/K)) < P(H)(1 – P(E/H)). This measure thereby favors theories with high information content cont(K) = 1- P(K) and high likelihood P(E/K). Following Hempel, these formulas are called measures of *systematic power*, when the capacities of a theory to give successful explanations and predictions are combined.

The simplest definition of *degrees of confirmation*, related to the High Probability criterion, is posterior probability:

$$\mathrm{conf}_1(\mathrm{H/E}) = \mathrm{P(H/E)}.$$

Popper's (1959) main criticism of probabilistic induction was that probability is a sign of logical weakness: if $\mathrm{H} \vdash \mathrm{K}$, then $\mathrm{P(H)} \leq \mathrm{P(K)}$. So science should not strive for probable hypotheses but for strong ones, which have high *information content* measured by

[18]For measures of explanatory and systematic power, see Hintikka (1968), Niiniluoto and Tuomela (1973), Festa (1999).

$$\text{cont(H)} = \text{P}(\sim \text{H}) = 1 - \text{P(H)}.$$

However, the Bayesians can point out that in science we are interested in high content backed up by evidence, so that one should combine the goals of low prior probability and high posterior probability.[19] This idea is to some extent reflected in the difference and ratio measures which are related to the Positive Relevance criterion[20]:

$$\text{conf}_2(\text{H/E}) = \text{P(H/E)} - \text{P(H)} = \text{P(H)}[\text{P(E/H)} - \text{P(E)}]/\text{P(E)}.$$
$$\text{conf}_3(\text{H/E}) = \frac{P(H/E)}{P(H)} = \frac{P(E/H)}{P(E)}.$$
$$\text{conf}_4(\text{H/E}) = \text{logP(H/E)} - \text{logP(H)}.$$
$$\text{conf}_5(\text{H/E}) = \text{logP(H/E)} - \text{logP(H/} \sim \text{E)}.$$

Measure conf_2(H/E) was proposed by Carnap (1962). All of these measures have a positive value if and only if E is positively relevant to H. They also satisfy the requirement that confirmation increases when the evidence is more surprising or improbable. One virtue of the difference measure conf_2(H/E), in comparison to the ratio measure conf_3(H/E), is that evidence gives strongest support to *a minimal explanation*, i.e. only to the part of an explanatory hypothesis H that is indispensable for the explanation of the evidence E. In other words, if H explains E, but A is irrelevant to E with respect to H, then conf_3(H/E) = conf_3(H&A/E) but conf_2(H/E) > conf_2(H&A/E), In this respect, the quantitative degree conf_2(H/E) corrects the qualitative conditions CE and CC which too generously attribute confirming power to non-minimal explanations.

Relating these measures of explanatory power and confirmation gives as results the following cases of (8). If H explains E better than K in the sense of expl_2, then conf_3(H/E) > conf_3(H/E) and conf_4(H/E) > conf_4(H/E). If H explains E better than K in the sense of the expected value of expl_1, then conf_2(H/E) > conf_2(K/E) (see Niiniluoto 1999a, 187). For many purposes, conf_2(H/E) gives similar results as the posterior probability measure conf_1(H/E). For example, we see that higher initial plausibility and higher explanatory likelihood together are sufficient for higher confirmation:

If P(K) $\geq$ P(H) and P(E/K) $\geq$ P(E/H) > P(E), then conf_2(K/E) $\geq$ conf_2(H/E) and conf_1(K/E) $\geq$ conf_1(H/E).

(Cf. Okasha 2000.) In particular, if evidence E is entailed by both H and K, then conf_1(H/E) > conf_1(K/E) iff conf_2(H/E) > conf_2(K/E) iff P(H) > P(K).On the other hand, there is an important difference between conf_2 and conf_3. Suppose that K is logically stronger than H, i.e. K entails H. Suppose also that K explains something

[19]See Hintikka and Suppes (1966), Levi (1967), Niiniluoto and Tuomela (1973).

[20]Measure conf_4(H/E) is defended by Milne (1996), conf_5(H/E) is proposed by Good in 1950 (see Good 1960). For surveys of confirmation measures, see Festa (1999), Kuipers (2000), who defends the ratio measure conf_3(H/E), and Fitelson (1999), who favors Good's measure conf_5(H/E).

(a part of the evidence E) that is not explained by H. Then K has greater explanatory power than H, but still K is less probable than H, i.e. P(K) < P(H), and P(K/E) < P (H/E) (see Monton 1998). However, in this situation it must be the case that P (E/K) > P(E/H), so that we may have $conf_2(K/E) > conf_2(H/E)$.

Peter Lipton (2004) distinguished "likely" explanations, which take into account the overall credibility or warrant of the explanation, and "lovely" explanations, which provide potential understanding about the explanandum. A likely explanation is probably true relative to the total evidence. He argued further that in many circumstances loveliness is at least a guide to likeliness. Lipton (2001b) concludes that likeliness can be defined by posterior probability P(H/E), whereas his notion of loveliness might correspond to explanations with a high value of P(E/H) (cf. Niiniluoto 1999b). Okasha (2000) interprets Lipton as saying that P (H) indicates likeliness and P(E/H) loveliness. But then it not clear why loveliness would generally be a guide to likeliness. Salmon (2001b) also considers the role of prior probabilities and likelihoods as possible explanatory virtues. However, the discussion remains largely inconclusive, since Salmon and Lipton do not appeal to any formal measures of explanatory power. They also seem to take it for granted that confirmation has to be explicated by high posterior probability – even though Salmon mentions the Popperian idea that scientists often prefer bolder hypotheses because of their informational value at the expense of their probability (*ibid.*, 121).

So far we have only discussed the relation of explanation and confirmation by arguing that the Bayesian model of inference helps to show how evidence may confirm hypotheses that are abductively introduced to explain them. However, the formal results about confirmation can be immediately generalized to more complex situations. First, all the probabilities can be relativized to background knowledge B. Secondly, besides the initial evidence E defining our original problem situation (with the explanation seeking question "why E?"), there may additional "old evidence" E_0, and it may happen that the hypothesis H proposed on the basis of E turns out to explain E_0 as well.[21] Then, using the difference measure $conf_2$, the confirmation of H by E and E_0 together is defined by $conf_2(H,E\&E_0)$, and the confirmation of H by E_0 relative to E by $conf_2(H,E_0/E) = P(H/E\&E_0) - P(H/E)$.

Thirdly, degrees of confirmation can be defined by similar formulas when new evidence is obtained. This evidence E' may come from surprising observations, which the available theory H is able to explain. (This was the case with the success story of Newton's mechanics, before it faced lacunae and anomalies.) But new predicted evidence E'' may also be the result of a test, performed after the adoption of H as a hypothesis from which E'' can be logically or probabilistically derived. In this way we obtain a probabilistic model of the method of hypothesis, eloquently discussed by Peirce in his later work. The prior plausibility of a hypothesis is now its probability P(H/E) relative to the initial evidence E, which depends on the ability of H to explain E. The confirmation of H due to the new evidence E' is then definable by $conf_2(H,E'/E)$. Again the joint confirmation of H by E and E' can be defined by

[21]Peirce was aware of this distinction, as one can see in *CP* 8.231–232.

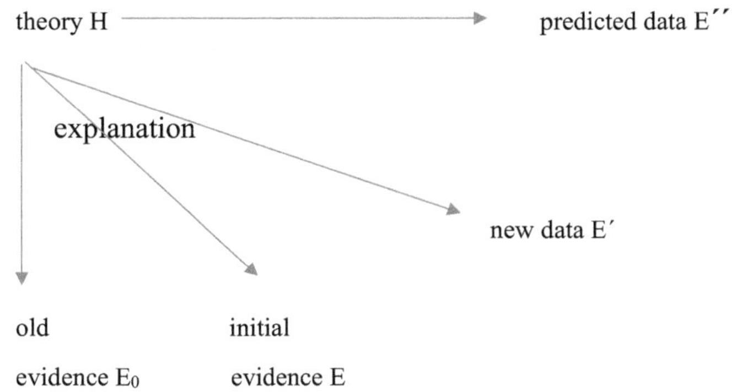

Fig. 6.1 The systematic power of a theory H with total evidence $E_t = E_0$ & E & E′& E″

$conf_2$(H/E&E′). The same applies to new predicted data E″. In typical cases, the novel facts E″ that H has predicted become, after the verification of E″, parts of the total evidence that H explains. Therefore, it seems that we need not assume an essential difference between accommodation and prediction as methods of confirmation (cf. Howson and Urbach 1989). Similarly, the measures of explanatory power can be reinterpreted so that they cover both explanatory and predictive relations between hypothesis and evidence. Hempel (1965) used the notion of *systematic power* as the combination of the explanatory and predictive power of a hypothesis relative to the total evidence E_t. Then the degree of confirmation of a hypothesis H on the total evidence E_t depends on the explanatory and predictive power of H with respect to E_t (see Fig. 6.1). For this reason, in discussions about abduction within the Bayesian framework we need not always distinguish between explanatory and predictive success, and rather be concerned with the relations between systematic power and confirmation.[22]

6.4 The Virtues of Unification

According to the traditional requirement, formulated already by William Whewell in his account of the "consilience of inductions" in 1840, an explanatory scientific theory should be independently testable by new kinds of phenomena. A good theory should have unifying power in the sense that it explains and predicts several mutually independent phenomena. This Section studies the prospects of Bayesianism to motivate this kind of unification criterion for abductive confirmation (see Niiniluoto 2008, 2016).

It is often required that a scientifically interesting hypothesis has to explain several mutually independent phenomena, thereby showing that these phenomena

[22]In spite of his doubts about Bayesianism, Psillos (2004), 90, finds this suggestion interesting.

are effects of a hypothetical common cause or a postulated unobservable entity (see Schurz 2008a).[23] This idea is historically related to the classical exposition of the method of hypothesis by William Whewell in *The Philosophy of the Inductive Sciences* in 1840. (Cf. Niiniluoto 1984, 37–38; Snyder 1997.) According to Whewell, a scientific hypothesis discovered by induction to account for phenomena should be independently testable by new kinds of predictions:

> The hypothesis which we accept ought to explain phenomena which we have observed. But they ought to do more than this: our hypothesis ought to *fortel* phenomena which have not yet been observed; – at least of the same kind as those which the hypothesis was invented to explain. ... But the evidence in favour of our induction is of a much higher and more forcible character when it enables us to explain and determine cases of *a kind different* from those which were contemplated in the formation of our hypothesis. (Whewell 1847, 62–65.)

Whewell argued further that the strongest sign of scientific progress occurs with "the consilience of inductions", where "an induction, obtained from one class of facts, coincides with an induction, obtained from another class". In consilience, inductions "jump together": two separate generalizations are found to be consequences of the same comprehensive theory. Peirce agreed with Whewell that scientific hypotheses, which are first inferred by abduction as explanations of surprising phenomena (*CP* 5.189), should be subjected to the test of experiment by deducing from them "most unlikely" experiental predictions (*CP* 7.182). Also Popper repeated Whewell's view in his requirement that a new theory should be independently testable: "it must have new and testable consequences (preferably consequences of new kind)" (Popper 1963, 241) (cf. Niiniluoto 1984, 37–38).

Whewell regarded consilience as "a test of the theory in which occurs", or "a criterion of reality, which has never yet been produced in favour of falsehood". Whewell seems to propose here an optimistic meta-induction from the history of science, claiming that the Peircean truth-frequency of consilience equals one. But this is clearly an exaggeration: a fallibilist has to admit that theoretical unification in science has led to mistakes as well.[24] The discovery of relativity theory and quantum theory in the early twentieth century shows that even Newton's long admired theory is not strictly speaking true but at best approximately true or truthlike (cf. Niiniluoto 1999a). How strong confirming power does such causal unification really have? How can we measure the strength of such confirmation?

One natural proposal is to regard a theoretical explanation the better, the more empirical phenomena it explains, and the less new entities or principles it postulates. For example, according to Philip Kitcher (1981), a theory achieves "unifying power" by "generating a large number of accepted sentences as the conclusions of acceptable arguments which instantiate a few stringent patterns" (cf. Schurz 1999; Morrison 2000). There is a trade-off between these desiderata, as new consequences may be achieved by adding assumptions and thereby making the theory more complex.

---

[23]The idea of common cause explanations was discussed in Chap. 4 in connection with natural sciences, medicine, and historical sciences (cf. Cleland 2002).

[24]For an estimate of the historical record of theoretical abductions, see Douven (2002).

This idea was formulated already by Eino Kaila in his Finnish monograph *Inhimillinen tieto* in 1939 with his notion of "relative simplicity".[25] According to Kaila, a theory has scientific value only to the extent that the multiplicity of its explanatory principles is smaller than the multiplicity of facts of experience that can be derived from it. Thus, the *relative simplicity* of theory H is the ratio between the multitude of empirical data E derivable from H and the number of logically independent basic assumptions of H.[26] In modern terms, this definition can be stated as the ratio between the explanatory power of H and the complexity of H.[27] Kaila, who had written in 1926 a monograph on *Wahrscheinlichkeitslogik* but had thereafter become sceptical about the possibility of quantitative inductive logic, further suggested that this ratio *would* be equal or at least proportional to the inductive probability of H given the data E, *if* it were measurable. His guess was not quite correct: when Hempel defined in 1948 the first formal measure of explanatory or systematic power of a theory H relative to evidence E, its value turned out to be P (~H/~ E) rather than P(H/E) (see $expl_1$ above)). But Kaila's important point was that relative simplicity in some way indicates the truth or epistemic acceptability of a theory, and is not only a pragmatic or conventional virtue in theory preference. Thus, his notion of relative simplicity can be viewed as a generalization of Whewell's consilience of inductions. For example, Newton's mechanics has high relative simplicity in Kaila's sense, since the law of gravitation explains many different kinds of phenomena (such as orbits of planets, free fall near the Earth, pendulum, etc.).[28]

With the formal machinery of epistemic probabilities, one can give a simple and straightforward argument about the Whewellian situation where theory H deductively explains two independent phenomena E and E′ (see Niiniluoto 2008).

Let E_1, E_2, ..., E_n be repeated occurrences of the phenomenon E, and let E(n) be their conjunction. Suppose that H logically entails E(n), so that P(E(n))/H) = 1. Then

[25]For an English translation, see Kaila (2014). Cf. Niiniluoto (1999a), 182.

[26]As the difficulties in the explication of the notion of complexity show (cf. Foster and Martin 1966), it is not easy to give a precise account of counting the number of the "logically independent basic assumptions" of a theory. This is illustrated by the debate of Myrvold (2003) and Schurz (2008b) on the comparison of the Ptolemaic and Copernican theories in astronomy.

[27]In the special of curve fitting, where the curve is defined by a function of order m and goes though n data points, the relative simplicity is the ratio n/m. Gillies (1989) proposes the difference n-m as a measure of "explanatory surplus". Currently the most popular approaches to curve fitting or "model selection" are the Akaike Information Criterion AIC and the Bayesian Information Criterion, which give different trade-offs between accuracy and simplicity (see Sober 2008, 82–104). When model M is a class of functions of order m, and L(M) is the member of M with the highest likelihood relative to data D, AIC gives log P(D/L(M)) – m as an unbiased estimate of the average predictive accuracy of M. Sober argues that this value is an unbiased estimate of the closeness to the truth of L (M), when closeness is measured by the directed divergence (8.15) (*ibid.*, 98).

[28]On the other hand, the religious hypothesis G-E (mentioned in footnote 14) would have minimal relative simplicity in Kaila's sense, since for each empirical statement E it requires the *ad hoc* premise that God wants E. Kitcher (1981) says that G-E achieves "spurious unification", but in fact it does not yield any kind of unification.

by (1) evidence E(n) confirms H. But repetitions of the same kind of evidence give diminishing returns, since by inductive learning the (n + 1)st occurrence of E is more probable than its earlier occurrences: $P(E_{n+1}/E(n)) > P(E_{n+1}) = P(E_1)$ (see Howson and Urbach 1989, 82). Now suppose further that H logically entails another phenomenon E′ which is probabilistically independent of E, i.e., $P(E'/H) = 1$ and $P(E'/E) = P(E')$. It follows from these conditions that $P(H/E(n)\&E_{n+1}) < P(H/E(n)\&E')$ iff $P(E') < P(E_{n+1}/E(n))$. The last condition holds under quite broad conditions, e.g., if the initial probabilities P(E) and P(E′) are about the same magnitude. Hence, relative to E(n), new kind of evidence E′ gives more confirmation to H than old kind of evidence E_{n+1}. This argument thus proves that successful explanation of the new kind of phenomenon E′ gives more confirmation to theory H than the repetition of the old kind of phenomenon E.

Another important result is proved by Wayne C. Myrvold (2003). Suppose that theory H achieves *inductive systematization* between empirical propositions E and E′ (cf. Niiniluoto and Tuomela 1973). This means that E and E′ are probabilistically independent, but they are informationally relevant to each other given H:

(IS) $P(E'/E) = P(E')$ and $1 > P(E'/E \ \& \ H) > P(E'/H)$.[29]

Myrvold proposes to measure the degree of *unification* of E and E′ achieved by H by means of the difference

(9) $$U(E, E'; H) = \log \frac{P(E'/E\&H)}{P(E'/H)} - \log \frac{P(E'/E)}{P(E')}.$$

It follows that U(E,E′;H) > 0 if H achieves inductive systematization between E and E′. Similarly, U(E,E′;H) > 0 if H achieves deductive systematization between independent E and E′ and H alone does not entail E′. Then Myrvold argues, by applying the logarithmic ratio measure of confirmation $conf_4$, that the degree of confirmation of H by E&E′ can be divided into three additive parts: confirmation of H by E, confirmation of H by E′, and the degree of unification of E and E′ achieved by H:

(10) $$conf_4(H/E \ \& \ E') = conf_4(H/E) + conf_4(H/E') + U(E, E'; H).$$

A similar result about the "virtue of unification" can be proved if $conf_4$ is replaced with Good's weight of evidence $conf_5$.

Essentially the same result is proved by Timothy McGrew (2003), 562 (cf. Schupbach 2005). Assume that H_1 and H_2 are equally good explanations of E and E′, i.e., $P(H_1) = P(H_2)$, $P(E/H_1) = P(E/H_2)$, and $P(E'/H_1) = P(E'/H_2)$.

---

[29]As an alternative to IS, one may require that H together with E is positively relevant to E′. i.e., $1 > P(E'/H\&E) > P(E')$. These notions of inductive systematization are motivated by Hempel's suggestion in 1958 that theoretical concepts could be logically indispensable for inductive systematization. Hempel had noted that deductive systematization by a theory H (i.e. $H\&E' \vdash E$ and not $E' \vdash E$) can always be achieved by an observational subtheory of H. As a way out of this "theoretician's dilemma", he proposed that inductive systematization might behave differently. Niiniluoto and Tuomela (1973) prove that Hempel's guess was right.

Assume further that H_1 achieves inductive systematization between E and E′ in the sense of (3), but E and E′ are independent conditional on H_2, i.e., $P(E/E'\&H_2) = P(E/H_2)$. Then a direct calculation by Bayes's Theorem shows that $P(H_1/E\&E') > P(H_2/E\&E')$. Hence, by all of our measures of confirmation, the theory H_1 which unifies E and E′ gains more confirmation by evidence E&E′ than the theory H_2 without such unifying power.

These results are related to the issue whether "coherence" is truth conducive (cf. Brössel 2013). The intuition that a coherent set of beliefs has a high probability of being true may arise from the controversial coherence theory of truth. Some Bayesians have given impossibility proofs against the proposal that coherence of evidence would increase the support for the hypothesis (see Bovens and Hartmann 2003; Olsson 2005). Tomoji Shogenji (2013) argues convincingly that some of the expectations of Bayesian coherentists have been wrong-headed, since it is the *diversity* of evidence that strengthens the support for the hypothesis. Shogenji's (1999) own measure of coherence for statements $E_1, \ldots, E_n$ is defined by

$$\mathrm{Coh}(E_1, \ldots, E_n) = P(E_1 \& \ldots \& E_n)/[P(E_1) \ldots P(E_n)],$$

and the conditional coherence of $E_1, \ldots, E_n$ given H is obtained by relativizing these probabilities to H:

$$\mathrm{Coh}(E_1, \ldots, E_n/H) = P(E_1 \& \ldots \& E_n/H)/[P(E_1/H) \ldots P(E_n/H)].$$

Now, applying the ratio measure of confirmation conf_3, we have

$$\mathrm{conf}_3(H/E_1 \& \ldots \& E_n) = [\mathrm{conf}_3(E_1/H) \ldots \mathrm{conf}_3(E_n/H)]\mathrm{Coh}(E_1, \ldots, E_n/H) \times /\mathrm{Coh}(E_1, \ldots, E_n).$$

But this equation can be rewritten as

(11) $\mathrm{conf}_3(H/E_1 \& \ldots \& E_n) = [\mathrm{conf}_3(H/E_1) \ldots \mathrm{conf}_3(H/E_n)]\mathrm{Coh}(E_1, \ldots, E_n/H)/\mathrm{Coh}(E_1, \ldots, E_n)$,

which shows that the confirmation H by $E_1, \ldots, E_n$ can be expressed as the product of the confirmation of H by each E_i, $i = 1, \ldots n$, and the increase of the coherence of $E_1, \ldots, E_n$ by H. By taking logarithm of both sides of (11), this equation reduces to Myrvold's result (10). In fact, the result (10) can be found already in Myrvold's earlier article (see Myrvold 1996, 663).[30]

Marc Lange (2004) has presented objections to Myrvold's treatment of unification (see also Schurz 2008b). With examples from physics and cosmology, he argues that Myrvold's unification is in some cases "too easy" to achieve. It is correct that formulas like (4) and (5) cannot reflect the question whether the unifying theory H is

[30]Olsson (2005) measures coherence by $\mathrm{Coh}^*(E_1, \ldots, E_n) = P(E_1 \& \ldots \& E_n)/P(E_1 \vee \ldots \vee E_n)$. This measure is related to J. Zamora Bonilla's (1996) epistemic measure of truthlikeness of H given E, defined by $\mathrm{Coh}^*(H,E)/P(H)$.

lawlike and explanatory or ontologically and causally prior to statements E and E′, and one could develop a more restricted notion of unification with these additional conditions (cf. (5) and (6)). But this does not yet discredit the Bayesian analysis of the virtue of unification.

Lange also points out that, for any independent E and E′, the simple conditional E → E′ unifies E and E′ in Myrvold's sense: U(E,E′;E → E′) > 0, since P(E′/E&H) = 1 and P(E′/H) is small. To avoid this technical objection, one can require that the unifying theory H should be logically independent of E′ and E. This is in fact required in the definition IS of inductive systematization.[31]

Myrvold (2003), 410, notes that on his account the conjunction E & E′ of two independent statements E and E′ does not unify them: U(E,E′;E&E′) = 0, since P(E/E&E′) = P(E′/E&E′) = 1. It is good that this kind trivial unification is excluded. But it may be somewhat surprising that, for similar reasons, Whewell's consilience of inductions is excluded as well (see *ibid.*, 418): if theory H entails both E and E′, which are independent of each other, then by (9) the degree of unification U(E,E′;H) is zero, since P(E′/E&H) = P(E′/H) = 1.[32]

Deductive consilience by Newton's theory does not show that e.g. Kepler's and Galileo's empirical laws are rendered informationally relevant to each other. On the contrary, these laws are probabilistically irrelevant to each other given the unifying theory.

A different probabilistic example with a similar of conclusion is given by Lange (2004). Let H be the hypothesis that Jones has a disease called systemic lupus erythematosus, and let E state that Jones has pleuritis and E′ that Jones has a malar rash. Lange specifies the probabilities so that pleuritis E and malar rush E′ are positively relevant to each other, but given lupus H, these symptoms E and E′ are independent of each other. Still, E and E′ are separately positively relevant to H and also jointly positively relevant to H, so that E&E′ strongly PR-confirms H. Indeed, using terminology from Hans Reichenbach's account of probabilistic causality, lupus is a *common cause* which screens off its probabilistic effects from each other. More generally, variable Z *screens off* variable X from variable Y iff X and Y are probabilistically dependent but become independent when conditionalized on Z (see Schurz 2008a, 344). Direct application of Myrvold's unification measure U to the values of X and Y given Z gives a negative value in such cases.[33] But, as we just saw, even in these cases a Bayesian treatment of the PR-confirmation of the common cause is available.

---

[31]See also footnote 9 on conditional deductive confirmation.

[32]Myrvold (2003) proposes to account for consilience by the idea that, in a law with free parameters, the values determined from one class of phenomena agree with those determined from another class. This is illustrated in William Harper's (2011) treatise on Newton's method.

[33]Schupbach (2005), who otherwise defends Myrworld against Lange, emphasizes the difference between Myrvold's unification and common cause explanations. However, in his reply to Lange, Myrvold (2017) argues that his measure U of mutual information unification applies to evidence statements which are sequences of values of X and Y, given a hypothesis which posits a common cause.

The outcome of our discussion is that there are two quite different concepts of unification, both of them legitimate as forms of abduction and with important applications in science (cf. Schurz 2015).[34] One of them is *linking up*: two empirical phenomena are independent from each other, but given a theory they become positively relevant to each other. This idea covers the notions of deductive and inductive systematization and Myrvold's measure of unification. The second is *screening off*: two empirical phenomena or variables are positively relevant or even indifferent to each other, but given a theory they become independent of each other. This idea covers Whewell's deductive account of consilience, Kaila's relative simplicity, and probabilistic common causes. In both cases, the unifying theory or common cause receives probabilistic confirmation from empirical phenomena.

[34]As the third relevant case one should mention examples of misleading similarities (convergence in evolution) and spurious correlations (e.g. between divorce rates and the consumption of margarine), where a common cause explanation is not plausible and should be replaced by separate cause explanations (see Sober 1988). But note that the search for such separate cause explanations is abductive as well.

Chapter 7
Inference to the Best Explanation

Abstract While Chap. 6 discussed the role of abduction in the confirmation of hypotheses by their success in explanation and prediction, in this chapter we turn to the notion acceptance which is a stronger form of justification than confirmation. Section 7.1 gives a survey of inductive acceptance rules, and following Gilbert Harman formulates inference to the best explanation (IBE) as a rule of acceptance: a hypothesis H may be inferred from evidence E when H is a better explanation of E than any other rival hypothesis. The notion of "best explanation" is explicated by measures of explanatory power, with a comparison to Lipton's distinction between "lovely" and "likely" explanations. In the special case with only one available explanation, IBE reduces to inference to the only explanation. Section 7.2 deals with the question of justifying IBE by giving replies to Bas van Fraassen's "bad lot" and "incoherence" arguments. It is concluded that under certain conditions an explanatory hypothesis may be so successful that its tentative acceptance as true is warranted.

In Chap. 6, we discussed the role of abduction in the confirmation of hypotheses by their success in explanation and prediction. In this Chapter, we turn to the notion acceptance which is a stronger form of justification than confirmation. In Sect. 7.1, inference to the best explanation (IBE) is formulated as a rule of acceptance. Section 7.2 deals with the question of justifying IBE, with replies to some important criticisms.

7.1 IBE as a Rule of Acceptance

The Bayesian treatment of inductive confirmation is sometimes interpreted as implying that after all induction is not a form of ampliative reasoning – so that Hume was right in rejecting induction as an inference. Bayes's Theorem (6.1) tells how new evidence transforms prior probabilities to posterior ones by means of conditionalization. This probabilistic reasoning is just mathematical calculation. Posterior probabilities allow the researcher to compare rival hypotheses for their

I. Niiniluoto, *Truth-Seeking by Abduction*, Synthese Library 400,
https://doi.org/10.1007/978-3-319-99157-3_7

credibility, but the influence of evidence does not lead even to the tentative acceptance of hypotheses. Howcver, when Bayesian posterior probabilities have been updated in "probability kinematics", they can be used in practical decision making for the calculation of the expected utilities of alternative actions. This view of Bayesianism was defended by Rudolf Carnap and among his followers Richard Jeffrey. Among statisticians, L. J. Savage argued that to "accept" a hypothesis means only a decision to act as if it were true.[1]

An alternative form of Bayesianism was defended among philosophers by Isaac Levi and Jaakko Hintikka. In his classic *Gambling with Truth* (1967), Levi contrasted his cognitivism with Savage's and Carnap's "behaviouralism" which seems to make scientists only advisers to decision-makers without results of their own. For a cognitivist, science seeks true information about reality, not only recommendations of practical action – event though the latter may be based on knowledge provided by basic research. According to Levi, induction has to be formulated by means of *rules of acceptance* which allow the researcher to tentatively add the best results of inquiry to the changing body of scientific knowledge. Such rules can be justified in cognitive decision theory, where the relevant *epistemic utilities* are taken to be truth and information and the acceptance of a hypothesis is defined by the maximization of expected epistemic utility. Similar views were developed by Hintikka and his students (see Hintikka and Suppes 1966; cf. Niiniluoto 2011c).

From the cognitivist point of view, results about positive *confirmation* do not yet guarantee that a hypothesis H would be *acceptable* on evidence E. The same evidence may weakly confirm several conflicting hypotheses, but this cannot be the case for acceptance. It may happen that no one of the rival hypotheses is clearly superior to others, so that suspension of judgment is the most rational conclusion. To improve the situation, new evidence is needed, or alternatively the problem has to be expanded by the introduction of new concepts. The extended cognitive problem contains then new rival explanations which may be evaluated by epistemic utilities. This version of cognitivism can be combined with fallibilism: when the best hypothesis so far has been found, it is always open to further revision and correction by new evidence. Even the anti-inductivist Popper allowed that the theories best corroborated by severe tests may "temporarily and tentatively, be accepted as part of current scientific teaching" (see Popper 1972, 261).

The lottery paradox shows that the High Probability Criterion of confirmation cannot define an inductive acceptance rule, i.e. high posterior probability is not sufficient for acceptance.[2] But in Hintikka's system of inductive logic high posterior probability is necessary for acceptance (see Niiniluoto 2011c). His framework

[1]Convergence theorems of de Finetti and Savage show that under some conditions researchers, who update their personal degrees of beliefs by the same evidence, are eventually led from different prior probabilities to the same posterior probabilities. See Howson and Urbach (1989) and Earman (1992). Objective Bayesians suggest that there are unique ways of fixing the prior probabilities.

[2]Similarly, low probability is not a sufficient condition for rejecting a hypothesis.

considers a classification system of K Carnapian Q-predicates, where each individual in the universe satisfies one and only one Q-predicate. Then prior probability is distributed to constituents C^w which specify which w Q-predicates are non-empty and which K – w Q-predicates are empty.[3] Thus, the logical form of a constituent is

(1) $(+/-)(Ex)Q_1x \;\&\; \ldots \;\&\; (+/-)(Ex)Q_Kx$,

where (+/-) is replaced by negation ¬ or nothing, so that C^w has K-w negation signs. Hintikka's typical assignment of prior probability satisfies $0 < P(C^w) \leq P(C^v)$ for $w \leq v$, whereas Carnap (1962) excludes all universal generalizations by assuming a priori that $P(C^K) = 1$. A Carnapian universe, where C^K holds so that all Q-predicates are instantiated, is atomistic in the sense that there are no true universal generalizations or laws. The evidence E consists of observations of n individuals which occupy $c < K$ Q-predicates. Such evidence falsifies some of the constituents, and with increasing n and fixed c all posterior probability tends to concentrate on one constituent C^c which claims that the universe is similar to the observed sample. The posterior probability P(H/E) of a universal generalization H, which is entailed by C^c, approaches one with increasing n, and when n is large enough H is acceptable on E (see Hilpinen 1968). But this inductive inference is fallible, since it is possible that the next individual in the sample falsifies C^c and H.

To apply the basic idea of cognitive decision theory, define an *ultimate partition* as a set $U = \{C_1, \ldots, C_n\}$ of mutually exclusive and jointly exhaustive statements in some language L, i.e. $\vdash \neg(C_i \;\&\; C_j)$ and $\vdash (C_1 \vee \ldots C_n)$ (see Levi 1967). Such partitions can be expanded by splitting its members e.g. by introducing new concepts to language L. In the simplest case, U contains a hypothesis H and its negation, i.e. {H,¬H}. An important example of an ultimate partition is the set of constituents of a first-order language. By its definition, an ultimate partition U contains a unique element C* which is its true member.[4] When C* is unknown, U defines a *cognitive problem* of finding C*, where the elements C_i of U are *complete answers* and disjunctions of elements are *partial answers* (see Niiniluoto 1987). The tautological disjunction of all elements $C_1 \vee \ldots C_n$ as a partial answer expresses ignorance ("I don't know") or suspension of judgment.

[3]For example, in a language with two monadic predicates F and G, the Q-predicates are $Q_1x = Fx \;\&\; Gx$, $Q_2x = Fx \;\&\; \neg Gx$, $Q_3x = \neg Fx \;\&\; Gx$ and $Q_4x = \neg Fx \;\&\; \neg Gx$. The generalization $g = (\forall x)(Fx \rightarrow Gx)$ states that the cell Q_2 is empty, so that it can be expressed as a disjunction of seven constituents which leave Q_2 empty. (The constituent which leaves all four Q-predicates empty is excluded, since the universe cannot be empty.) If the probabilities of finding an individual b in the four Q-predicates are equal, then we have $1/3 = P(Fb/g) < P(Fb/Gb \;\&\; g) = ½$, which shows that Gb confirms Fb relative to g (cf. (6.5)). But to formalize singular abduction (1.3), so that Fb is acceptable on Gb and g, we need to assume that the probabilities the Q-predicates are non-symmetric, i.e. $P(Q_1b)$ is much larger than $P(Q_3b)$.

[4]This means that in inferential problems, which are defined relative to an ultimate partition, there is no problem with the "bad lot" (cf. van Fraassen, 1989). See below Section 7.2.

Let u(H,t,E) be the epistemic utility of accepting hypothesis (partial answer) H on evidence E, when H is true, and u(H,f,E) the epistemic utility of accepting H on E, when H is false. Then the *expected epistemic utility* of accepting H on E is

(2) $U(H, E) = P(H/E)u(H, t., E) + P(\sim H/E)u(H, f, E)$.

The epistemic acceptance rule can now be defined by

(EA) Accept on evidence E the hypothesis H which maximizes the value U(H,E).

If we are interested in truth and nothing but the truth, then a natural choice for the epistemic utilities would be u(H,t,E) = 1 and u(H,f,E) = 0, i.e. the truth value of H. But then U(H/E) = P(H/E), which has the maximum value one when $E \vdash H$, so that the rule EA would lead to the conservative view of accepting only the logical consequences of evidence. In Popper's spirit, Levi's (1967) solution is to combine our interest for truth with a "relief from agnosticism", so that we have to "gamble with truth" by making our epistemic utility a weighted combination of a truth-factor and an information factor. Therefore, a rule of acceptance should include an "index of boldness" which tells how willing the scientist is to risk error in the attempt to relieve from agnosticism. Hintikka's modification of this idea to choose u(H,t,E) = cont(H) and u(H,f,E) = - cont(~H), i.e. in accepting a false hypothesis we lose the content of its negation. This leads to the expected utility

(3) $U(H, E) = P(H/E) - P(H)$

which equals the difference measure of confirmation $conf_2(H/E)$ of Sect. 6.3 (see Hintikka and Suppes 1966). Note that (3) is equal to P(H/E) + cont(H) – 1. Like Levi's measure, it thus favors hypotheses which combine high information content with high posterior probability P(H/E). If the acceptance rule based on (3) is formulated by following Levi's model with ultimate partitions, it recommends the rejection of all elements C_i with $P(C_i/E) < P(C_i)$ and the acceptance of the disjunction of all unrejected elements as the strongest on the basis of E.[5] Assuming that inductive explanation is explicated by the criterion positive relevance (cf. Sect. 6.2), this means that all complete answers C_i which are not explanations of E are ruled out, and the disjunction of all explanatory answers C_i is accepted as the strongest on E. Another selective way of using (3) is to apply it for the comparison of the explanatory complete answers and choosing the best among them.[6]

[5]In Risto Hilpinen's (1968) modified version, the expected utility is P(H/E) – qP(H), where $0 < q \leq 1$ is an "index of boldness" in Levi's sense. Thus, for (3), this index of boldness is maximal. However, even the boldest application of Levi's rule is cautious in the sense that it always prefers truth to falsity. In Chap. 8, we shall introduce measures of truthlikeness which admit that some false statements may be so close to the truth that they are cognitively better than weak uninformative truths like tautologies.

[6]Levi (1979) restricts the task of abduction to the formulation of potential hypotheses, without reference to explanation, but here we are interested in the role of abduction in the selection of the best hypothesis.

The cognitivist approach to induction suggests that the theory of abductive reasoning should not treat the probabilistic results of Chap. 6 on abductive confirmation as the final word about scientific knowledge. One should also ask whether abduction can be formulated as a rule of acceptance. We have seen in Chap. 1 that Peirce himself made hints in this direction, even though he knew that abductive reasoning is not logically valid. He was aware that in some cases inference from effect to cause may be irresistible or compelling "like a flash". This happens in perceptual judgments which are "an extreme case of abductive inferences" (*CP* 5.181-185). Thus, unless we accept phenomenalism which denies the causal element in perception, we have to acknowledge that our everyday knowledge about the observational properties of ordinary physical objects (such as stones and tables) involves abductive inference. When sitting in my room I hear a rambling sound outside, I immediately infer that there is a car in the street, excluding thereby alternative implausible explanations like hallucinations. Another important area of human knowledge which is based upon inverse reasoning or retroduction is history (*CP* 2.714). In these examples about perceptual and historical judgments, hypothetic inference for Peirce is not only a method of discovery but also a fallible way of *justifying* an explanation (see also Chap. 4). We have also seen in Chap. 3 that many logicians and AI researchers formulate abduction as a rule of inference, which amounts to treating it as a rule of acceptance.

Hempel required in his early papers that an adequate explanation should be true (see Hempel 1965, 246). For a scientific realist, this is a natural desideratum. For example, when we explain the death of a person, the best explanatory account should identify the *real cause* of the explanandum event. However, in science we usually cannot directly verify the truth of an explanatory hypothesis, but at best we can make indirect inferences and tests about its truth value. It would be ineffective to formulate abduction as the rule to infer the *true* explanation from the given data. Therefore, as Hempel recognized, we need also the concept of potential explanation, and then it is important to ask whether the ability of a hypothesis to serve as a potential explanation of known facts in any way indicates its truth. In Chap. 6 we already saw that one explanation may be *better* than another. Using these criteria for comparing explanations, the natural next step is ask which hypothesis is the *best* explanation of the available empirical data.

Thus, in the strong interpretation, abduction is not only an inference to a potential and perhaps testworthy explanation but to the best explanation. According to Gilbert Harman (1965), *inference to the best explanation* (IBE, for short) recommends the inference from evidence E to hypothesis H when H is a better explanation of E than its rivals:

(IBE) A hypothesis H may be inferred from evidence E when H is a better explanation of E than any other rival hypothesis.

Here inference is understood in the strong sense of acceptance as true. IBE thus claims that the best explanation H of evidence E is rationally acceptable on E:

(IBE′) If hypothesis H is the best explanation of evidence E, then conclude for the time being that H is true.

Harman also tries to argue that all inductive inference can be reduced to IBE.[7]

A typical formulation of IBE in the AI literature is the following:

D is a collection of data (facts, observations, givens)
H explains D (H would, if true, explain D)
No other hypothesis can explain D as well as H.
Therefore, H is probably true.

Here the criteria of the best explanation include: how good H is by itself, how decisively H surpasses the alternatives, and how thorough was the search for alternative explanations (see Josephson and Josephson 1994). The term "probably" means that the conclusion H does not follow deductively from the premises (see Psillos 2009, 183, 212).[8]

The notion of "the best explanation" is of course ambiguous in several ways. For example, it relevant to distinguish *deductive* explanations, where the explanadum E is deducible from the explanans H, and *inductive-probabilistic* explanations, where the explanandum follows from the premises only with some probability. The relativity of such inferential conditions from H to E the background knowledge B means that the converse abductive inferences from E to H are context-dependent (cf. Day and Kincaid 1994). Particularly interesting are cases, where the background information B together with evidence E excludes all potential explanations but one. Here exclusion may be understood in the pragmatic sense that B and E give us good reasons to reject some of the available (so far discovered) explanations. So as a special case of IBE, where a hypothesis has no serious rivals, we have *inference to the only explanation*:

(IOE) A hypothesis H may be inferred from evidence E when H is the only available explanation of E.

In the limiting case, which Bird (2010) calls "eliminative abduction", evidence eliminates by Modus Tollens all but one of the hypotheses. If the list of hypotheses is exhaustive, so that the conditions of an ultimate partition are satisfied (cf. Sect. 6.2), then such eliminative abduction is a deductively valid argument.[9]

An important distinction can be made between local and global theorizing (see Hintikka 1968). In *local* theorizing, we are interested in finding an explanation of a

[7]For discussion of this thesis, see Sect. 1.4.

[8]See also formulations of abduction as a rule of inference in Chap. 3.

[9]Bird (2010), who calls such deduction "Holmesian inference", argues that Lipton's (1991) historical example of Ignaz Semmelweis's study of childbed fever (see Hempel, 1966) is an eliminative abduction in this strong sense. But one may doubt that the list of potential explanations that Semmelweis considered was exhaustive.

particular piece of data or evidence E, while in *global* theorizing evidence E is only a stepping-stone to a more generally applicable theory. Hintikka suggests that in the local case it is natural to look for a theory H that transmits a maximal amount of information about the given evidence E. If semantic information is defined by a logarithmic measure of "surprise value" –logP(E/H), this demand leads to the principle of *Maximum Likelihood*:

(4) Given evidence E, accept the theory H which maximizes the likelihood P (E/H).

The same rule follows if the value of an explanation H with respect to evidence E is measured by expl_2(H,E). This rule is a standard method in statistical estimation problems.

For the case of deductive explanation, the rule (4) does not help much, since it cannot distinguish between potential explanations that make the likelihood P(E/H) equal to one. But even here a difference in the rival hypotheses may be detected by enlargening the description of the evidence E. For example, suppose we are looking for an explanation of the death of a certain person, and there are available several possible causes H, H′, ... of her death. However, if the case is given a fuller description, including facts about her life and the symptoms in her body, it may be that many of the potential hypotheses fail to give a causal explanation any more. In the limit, it may happen that only *one* of the potential hypotheses is left, and this is by IOE certainly the "best" explanation in this situation.[10]

In global theorizing, we are trying to find a good hypothesis H which explains also other phenomena than the particular data E. Here it may seem natural to take the amount of the substantial *information content* of H, measured by cont(H) = 1 - P(H), as the epistemic utility to be maximized:

(5) Given evidence E, accept the theory H which maximizes the information content cont(H).

However, this rule has the undesirable feature that it favors a logically inconsistent theory H (with P(H) = 0 and cont(H) = 1). Moreover, it always prefers a logically stronger theory to a weaker one, even when the additions are irrelevant to the evidence E. It does not help at all to replace cont(H) by the relative content 1 – P (H/E) of H given E (see Hintikka 1968). Hempel's measure of the systematic power expl_1(H,E) = P(~H/~E) of H with respect to E is maximized by minimizing P(H) (1 – P(E/H)), so that it favors hypotheses with high content and high likelihood, but it also receives its maximal value for a contradiction H. Hence, the rule IBE leads to absurd consequences if the "best explanation" is simply characterized as the theory which has the maximal degree of information content or systematic power.[11]

---

[10]If the coroner is asked to identify *the* cause of death, a disjunction of even two alternative causes may be too weak.

[11]It is questionable whether IBE can be saved from these troubles by joining it with simplicity considerations, such as "consilience" in Thagard (1978) and "explanatory unification" in Friedman (1974) or with additional penalties for "conceptual problems" (cf. problem-solving ability in Laudan, 1977).

The situation changes decisively if information content and systematic power are treated as *truth-dependent* epistemic utilities, as shown by Levi, Hintikka and Juhani Pietarinen (see Hintikka 1968; Niiniluoto 1990a, 1999a, 187). For example, if the utility of accepting H is cont(H) when H is true and -cont(~H) when H is false, then by the result (3) the rule (5) is replaced by the recommendation to accept the hypothesis with the highest value of the relevance measure of confirmation:

(6) Given evidence E, accept the theory H which maximizes P(H/E) - P(H).

The same result is obtained by choosing the two utilities as $\mathrm{expl}_1(\mathrm{T,E})$ and $-\mathrm{expl}_1(\sim\mathrm{H,E})$. According to this rule, the best hypothesis should have both high posterior probability given evidence and high information content. As posterior probability is equal to the expected truth-value of H given E, the rule (6) is also obtained by choosing truth and information as the basic utilities. Thus, (6) is a way of balancing the demands of truth and information (systematic power) in scientific inquiry.

If hypothesis H is logically stronger than hypothesis H′, i.e. H entails H′, then we have $P(H) \leq P(H')$ and $\mathrm{cont}(H) \geq \mathrm{cont}(H')$. If, in addition, both H and H′ entail a contingent evidence statement E, then we have $P(H/E) - P(H) \leq P(H'/E) - P(E)$. In this deductive case, the rule (5) recommends that a logically weaker hypothesis is always better than a stronger one. However, in the probabilistic case, where the hypotheses do not entail the evidence, this need not be the case. An example is sketched by Nelson (1996), 404. More generally, in Hintikka's logic of inductive generalization the difference in rule (6) is maximized by the logically strongest generalization C^c compatible with a sufficiently large evidence.

Peter Lipton (1991) suggests that IBE should be understood as "Inference to the Loveliest Potential Explanation": a *lovely* explanation is one which offers a high degree of "potential understanding". Lipton points out that the loveliest explanation is not always the same as the "likeliest": a *likely* explanation is probably true relative to the total evidence. (Note that in this terminology the Maximum Likelihood principle (4) is a local loveliness condition.) But, as a reply to "Voltaire's objection" about the correlation of loveliness and truth, he argues that at least on some conditions the loveliest explanation is also the likeliest potential explanation. Lipton, who does not formalize his argument, is criticized by Eric Barnes (1995). A more technical argument to the effect that epistemically true theories and "best-explaining theories" coincide has been given by Raimo Tuomela (1985).

Search for "likely" explanations would be justified by the high probability rule which recommends the acceptance of the most probable explanation:

(7) Given evidence E, accept the explanation H of E such that H has the maximal posterior probability P(H/E) on E.

As we noted, this rule can be justified by taking truth as the only epistemic utility. As self-explanations are excluded, in (7) H cannot be E (or contain E as its conjunct), even though P(E/E) has the maximal value 1. The condition that H has to explain E also excludes trivial tautologies as candidates for H. For theories which deductively explain E, it follows from Bayes's Theorem that rule (7) recommends the choice of the theory H that has the highest prior probability P(H), i.e. H is initially the most plausible of the rival explanations. With the same assumptions, this H also maximizes the difference P(H/E) - P(H), so that (6) and (7) are equivalent in this case. However, as we already noted, if rule (6) is applied to inductive explanation as well, (6) and (7) may lead to different results: the theory with maximal likelihood need not be the one with the highest posterior probability, if it has a very low prior probability. The criticism that the likelihood principle (4) commits the "base rate fallacy", as it ignores the prior probabilities, is usually based on the idea that abduction should lead to hypotheses with high posterior probability (see Psillos 2004, 2009, 56).[12]

Besides truth, information, and systematic power, *simplicity* is often proposed as a criterion for the choice of theoretical hypotheses in science (see Sober 2015). The medieval Ockham's Razor famously recommended parsimony as the criterion of satisfactory ontology. If an instrumentalist favors simplicity as a replacement of the virtues of theoretical explanatory power (e.g. demanding that we choose the simplest hypothesis compatible with data), we may enter into a conflict with the realist idea of abduction. But as Peirce himself argued, simplicity (or "incomplexity") could guide us in choosing between alternative explanations. Especially in the context of pursuit (see Sect. 5.3) it makes sense to choose the simplest theory as the object of further inquiry, testing, and elaboration. This is one interpretation of Hans Reichenbach's (1938) principle of inductive simplicity (see Niiniluoto 1994, 1999a, 179–184). But in the context of justification or acceptance, it is not generally clear that the simplest theory would be the best candidate for truth. For example, biological evolution does not always follow the most parsimonious paths (cf. Sect. 4.4). The old dictum *simplex sigillum veri* (i.e. simplicity is a sign of truth) is sometimes defended by assuming that greater simplicity is associated with higher prior probability (Salmon 1990), so that in the deductive case the simplest explanation also has highest posterior probability. On the other hand, in Hintikka's system of inductive logic the most parsimonious constituent compatible with evidence (i.e. C^c) will be the most probable a posteriori for sufficiently large evidence.

We argued above in Sect. 6.3 that explanation and prediction behave in a symmetrical fashion with respect to the probabilistic confirmation of rival hypotheses. However, this is problematic in the case of acceptance. A hypothesis that explains our initial data, and is thereby confirmed by it to some extent, may still be ad hoc. To remove this doubt, the hypothesis should be independently testable, i.e. it should either explain some old evidence or be successful in serious new tests. As we have seen in Sect. 6.1, this is what Peirce (following Whewell) demanded. Therefore, one may argue that IBE as an acceptance rule should contain an additional condition stating that the "best" hypothesis is one with both explanatory and predictive power. For example, the Standard Theory will not be accepted by its

[12]For an attempt to find objective prior probabilities for IBE, see Weisberg (2009). In Carnap's and Hintikka's systems of inductive logic, there are canonical ways of giving prior distributions and likelihoods, but they are open up to the choice of one or two parameters.

great explanatory power alone, but it has to be successful in experimental tests.[13] Historical hypotheses about past singular events are primarily assessed by their explanatory power (see Sect. 4.1), but they too should be able to give an account of some novel effects.

As a rule of acceptance, IBE also faces the problem that sometimes the best explanation is not good enough for being a serious candidate for truth (see Salmon 2001a, 85–86), even though (as we have seen) the evidence gives some confirmation of it. Lipton (2001b, 104), suggests that IBE should be formulated as "inference to the best of the available competing explanations, when the best one is sufficiently good". But then we need a criterion for saying when an explanatory hypothesis is "good enough" for acceptance (cf. Douven 2011). The answer can be sought from the quantitative measures of explanatory power: a hypothesis H is sufficiently good if its degree of explanatory power expl(H,E) with respect to evidence E exceeds a threshold value. Depending on the choice of expl, this gives a constraint on the likelihood P(E/H), the difference P(H/E) – P(H), or the posterior probability P(H/E). But note that if the Bayesian account of acceptance allows for the strongest alternative to be a disjunction of the members of an ultimate partition (cf. Levi 1967), then depending on the boldness of the inquirer the best explanation may be a single complete answer or a disjunction of some alternative explanatory answers. In some cases *suspension of judgment* is the most rational attitude towards rival hypotheses. In practice, such temporary suspension of judgment can be interpreted as the recommendation that the rival explanations are still pursued - developed theoretically and tested experimentally - until one of them is deemed good enough for tentative acceptance (cf. Sect. 5.3). In lucky cases, one may find new evidence which as a "smoking gun" unambiguously discriminates one of the available hypotheses as the best explanation of the observed facts (cf. Cleland 2002).

7.2 Replies to Criticism

Bas van Fraassen's (1980, 1989) much discussed criticism of abduction concerns IBE as a rule of acceptance, i.e. as a rule that "selects" the best among a set of proposed explanations.[14] His discussion of the possibility of a Bayesian analysis of IBE operates with the notion of belief related to high posterior probability, so that his

[13]An important contemporary example of the power of explanatory considerations is the postulate of dark matter. The majority of physicist accept the existence of dark matter as an abductive solution to the problem of "missing masses", since it explains anomalous observations about galaxies, even though dark matter by definition is invisible (does not interact with electromagnetic radiation), and experimental attempts to verify or test its existence have failed. Therefore, some philosophers have claimed that it is an untestable ad hoc hypothesis (see Merritt, 2017). See also Sect. 8.5.

[14]For critical evaluation of van Fraassen's arguments, see Kvanvig (1994), Psillos (1996, 1999), Douven (1999), Okasha (2000), Niiniluoto (2004), and Henderson (2014). For defense, see Ladyman et al. (1997).

target can be taken to be (6). However, his three arguments, if they were valid, would also serve to refute the connection between abduction and confirmation.

Peirce repeatedly appealed to Galileo's notion of *il lume naturale*, our natural disposition to light upon the correct hypothesis in a finite number of guesses (*CP* 7.223). According to Peirce, "man has a certain Insight, not strong enough to be oftener right than wrong, but strong enough not to be overwhelmingly more often wrong than right" (*EP* 2:217). But as IBE is always restricted to a set of historically given or formulated hypotheses, it may lead to "the best of a bad lot". How could we know that the true hypothesis is among the so far proposed? And as this set is in any case very large, its "random member" must be quite improbable. A simple Bayesian reply to this is to point out that Bayes's Theorem has to be applied to an ultimate partition, consisting of a set of mutually exclusive and jointly exhaustive hypotheses, so that the disjunction of hypotheses $H_1, ..., H_n$ is a logical truth and, hence, precisely one of them is true (see Niiniluoto 1999b).

More technically, van Fraassen's *bad lot argument* claims further that IBE does not satisfy even the minimum condition that an acceptable hypothesis should have at least the probability 1/2:

> For me to take it that the best of set X will be more likely to be true than not, requires a prior belief that the truth is already more likely to be found in X, than not. (Van Fraassen 1989, 143.)

But this claim is in fact incorrect. To see this, assume that $n = 3$, H_1 deductively explains E, and H_2 inductively explains E, but $p_1 + p_2 < .5$, so that our prior belief is that the truth is not to be found in the set $\{H_1, H_2\}$. For example, let $p_1 = p_2 = .2$, $p_3 = .6$, $q_1 = 1$, $q_2 = .6$, and $q_3 = .1$. Then the best explanation of E is H_1. By Bayes's Theorem (6.1), $P(E) = .38$, and the posterior probability of H_1 on E is .53.

More interesting issues can be raised about the relativization to an ultimate partition. As Shimony (1970) pointed out, if we have say two serious hypotheses H_1 and H_2, then an ultimate partition can be formed with a third "catch-all hypothesis" $\neg(H_1 \vee H_2)$. But, as Douven (2011) points out, such catch-all hypothesis need not be an interesting hypothesis – or an explanation at all. Kyle Stanford (2006) has continued van Fraassen's bad lot argument by arguing that even the choice from a "very good lot" by IBE still faces the worry that "there may be other, equally good alternatives which remain presently unconceived" (*ibid.*, 48), since recurrently the scientists have failed to exhaust all the possible hypotheses. But in my view the potential existence of "unconceived alternatives", which may be equally well supported as our currently best one, need not lead to a theoretical scepticism about atoms, genes, gravitational fields, and the past ancestors of our species (*ibid.*, 210). On the contrary, the realist need not believe that our current descriptions of theoretical entities are final and literally true, since progress can still be made with the advance of theories and evidence for them. Such improvements of our so far best theories can be found by expanding an ultimate partition – either by deepening the best hypothesis with a stronger revision or by splitting the catch-all hypothesis into interesting alternatives. In this sense, the fact that truth has not yet been found gives hope for optimism: perhaps still better and deeper theories can be reached when we move to new and richer conceptual frameworks (cf. Niiniluoto 2017).

We have seen that theoretical hypotheses H with P(H) > 0 may receive support from their empirical successes, so that their posterior probability P(H/E) may be high. But the application of formula (6.2) for P(E) presupposes that the probability of evidence E on all alternative hypotheses H_i is known. If H_1 is the explanation of E, then we should know P(E/¬H_1), which is the weighted average of the likelihoods P (E/H_i) for $i \neq 1$. It was argued already by George Boole in 1854 that this assumption is unrealistic (see Boole 1958, 367; cf. Salmon 1990). However, if P(E/H) = 1 and if we allow P(E/¬H) to take any values in the open interval (0,1), then by (6.1) P(H/E) takes values in the interval (P(H),1). This is another way stating the Basic Theorem (6.5): successful deductive explanation of E by H increases the probability of H on E, even when the likelihood P(E/¬H) is unknown. If this likelihood is sufficiently small, so that H is the only explanation of E, then the increase in the probability P (H/E) will be significant. In the limiting case of IOE, where P(E/¬H) = 0, we have P (H/E) = 1 independently of the prior probability P(H) > 0 (cf. Psillos 2009, 59).[15]

Van Fraassen's second criticism claims that the Bayesian account of abduction leads to incoherence: if explanatory success is credited with adding an extra bonus to the posterior probability, this procedure allows the construction of a Dutch Book against the scientist (*ibid.*, 169). However, the point of the Bayesian approach is not to add separate bonus points, but to show that explanatory virtues enter into the assessment of posterior probability via the normal elements of Bayes's Theorem, viz. prior probability and likelihood (cf. Kvanvig 1994). No such special assumption was needed to prove the basic result (6.5) about the connection between explanation and confirmation.

Van Fraassen's third criticism notes that the result (6.5) about confirmation is blocked if the prior probability P(H) is zero. This is the most interesting of his objections to IBE, since the reply needs new logical tools - the notions of approximate truth and truthlikeness - beyond the probability calculus. In the next Sect. 8.3 we shall discuss two cases in which such a zero probability is natural: (a) H is a *sharp* hypothesis, i.e. H belongs to an infinite set of rival hypotheses **H**, e.g. the set of real numbers; (b) H is an *idealized* hypothesis, i.e. the class of rival hypotheses (including H) is defined relative some counterfactual assumption (cf. Niiniluoto 1999a). In Chap. 9 we try to rebut the objection that the use of abduction to defend scientific realism is question-begging or circular (cf. Laudan 1984; Fine 1986; van Fraassen 1989).

In this chapter, we have analyzed IBE as a rule of acceptance within the Bayesian probabilistic framework. But this is not the only way of looking at ampliative inferences which aim to be epistemically probabitive. If IBE can be defended independently of probabilities, van Fraassen's critique could be dismissed as irrelevant. An interesting attempt in this direction is given by Stathis Psillos (2002, 2009) by employing the notion of defeasibility. Ampliative non-deductive forms of reasoning like induction and abduction are defeasible, i.e. liable to counterexamples or

[15]Another case which guarantees that P(H/E) is close to one is that H has a unique novel deductive prediction E which no rival of H is able to predict (see Leplin 2004).

"defeaters", but their conclusions are epistemically warranted if the possible defeaters are not actual. In the case of IBE such defeaters could be hitherto unthought but better explanations. Psillos appeals to coherence conditions (cf. Thagard 1989) and contextual background knowledge to show that IBE "has resources to show when a potential undercutting defeater can be neutralized". Even if IBE is ampliative, and cannot be reduced to a conclusive deductive argument, "one cannot seriously question the fact that a hypothesis stands out as the best explanation of the evidence offers defeasible reason to warrantedly accept this hypothesis" (Psillos 2009, 190).

Chapter 8
Abduction and Truthlikeness

Abstract Earlier chapters deal with abductive inferences to explanations which are deductive or inductive-probabilistic. This more or less standard account has so far ignored the fact that explanatory and predictive success in science is often approximate. Therefore, the analysis of abduction should cover also approximate explanations, which is illustrated by Newton's explanation of Kepler's harmonic law (Sect. 8.1). The notions of approximate truth (closeness to being true), verisimilitude (closeness to complete qualitative or quantitative truth) and legisimilitude (closeness to the true law) are defined in Sect. 8.2. This leads us to generalize Peirce's model of abduction to cases where the conclusion states that the best theory is truthlike or approximately true, with illustrations from idealized theories and models (Sect. 8.3). In a comparative formulation, if theory Y is a better explanation of the available evidence E than theory X, then conclude for the time being that Y is more truthlike than X. To justify such abduction, we need a method of estimating degrees of truthlikeness by their expected values. Another tool is the notion of probable approximate truth. Then, in order to answer to Laudan's challenge, the probabilistic link between empirical success and truth has to be replaced with a fallible bridge from the approximate empirical success of a theory to its truthlikeness (Sect. 8.4). Section 8.5 gives some remarks on abductive belief revision, which is related to cases where the evidence is conflict with the theory. This theme extends Aliseda's way of linking belief revision models with abductive reasoning.

In the earlier Chapters we have discussed abductive inferences to explanations which are deductive or inductive-probabilistic. This more or less standard account has so far ignored the fact that explanatory and predictive success in science is often approximate. Therefore, the analysis of abduction should cover also approximate explanations (Sect. 8.1). The notions of verisimilitude and legisimilitude are defined in Sect. 8.2. This leads us to generalize Peirce's model of abduction to cases where the conclusion states that the best theory is truthlike or approximately true (Sect. 8.3). To justify such abduction, we need a method of estimating degrees of truthlikeness, and the probabilistic link between empirical success and truth has to

I. Niiniluoto, *Truth-Seeking by Abduction*, Synthese Library 400,
https://doi.org/10.1007/978-3-319-99157-3_8

be replaced with a fallible bridge from the approximate empirical success of a theory to its truthlikeness (Sect. 8.4). The concluding Sect. 8.5 gives some remarks on abductive belief revision.

8.1 Approximate Explanation

The notion of approximation has an important function in science. Charles S. Peirce, in his Harvard lecture in 1903, suggested that in assessing "the agreement of the theory with fact" we need a "provisional and approximative, or probametric measure" (*EP* 2:225). In the draft, he explained that "probametric" is a single word for the more clumsy "probable and approximate", which he associated with the idea of "probable error" (*EP* 2:530).[1] In this Section, the approximation of laws is illustrated by Kepler's harmonic law. Measures for the distance between laws are introduced in the next Section.[2]

Johannes Kepler discovered his first two laws about planets in 1609. The first law states that the orbit of each planet is an ellipse with the Sun at one focus. The second law states that a line joining a planet and the Sun sweeps out equal areas in equal times. Kepler's Third Law was discovered 10 years later. According to this Harmonic Law, the square of the orbital period of a planet (t) is directly proportional to the cube of the semi-major axis of the planet (a), i.e.

(1) $t^2/a^3 =$ constant for all planets.

Here a is equal to the average distance of the planet from the Sun.

For the empirical data available to Kepler, the law (1) was remarkably accurate. When t is given in years and a in astronomical units (a = 1 for the Earth), he could make the following calculations:

	t	a	t^2/a^3
Mercury	0.241	0.3881	0.994
Venus	0.616	0.7240	1.000
Earth	1.000	1.000	1.000
Mars	1.880	1.5235	0.9995
Jupiter	11.863	5.1965	1.0029
Saturn	29.458	9.5100	1.0089

[1]Sloughter (1996) argues that plausibility in Peirce's probabilistic abduction (see e.g. (1.9)) can be measured by the significance level in R. A. Fisher's theory of statistics. Significance level tells how unlikely the evidence must be for the null hypothesis to be rejected, so that in a sense the improbability of data given a null hypothesis measures their discrepancy. This idea is related to the likelihood principle (7.3). But Slaughter does not discuss the controversial aspects of Fisher's "fiducial" inference, which attempts to reach something like doxastic probabilities without Bayesian priors.

[2]This Chapter is partly based on Niiniluoto (2005a, 2014b).

The law (1) further predicts fairly accurate values for the later discovered planets Uranus and Neptune.

Newton's mechanics in *Principia* in 1687 was celebrated for its ability to explain Galileo's and Kepler's empirical laws. William Whewell praised in 1840 the unifying and simplifying power of Newton's theory which explained all of Kepler's three laws, "although no connexion of these laws had been visible before", and also surprisingly accounted for the fact known as the precession of the equinoxes (cf. Sect. 6.4). Also empiricist philosophers, like John Stuart Mill in his *A System of Logic* (1843), accepted the explanation of laws as an important task of science. However, in 1907 the French instrumentalist Pierre Duhem, who feared that the aim of explanation would subordinate science to metaphysics, criticized the Newtonian method:

> The principle of universal gravity, very far from being derivable by generalization and induction from the observational laws of Kepler, formally contradicts these laws. If Newton's theory is correct, Kepler's laws are necessarily false. (Duhem 1954, 193.)

Karl Popper welcomed this argument in his campaign against induction. In a lecture in 1948 he pointed out that "Kepler's laws cannot be deduced from Newton's", but can be obtained only *approximately* by making "the assumption that masses of the various planets are negligible compared with the mass of the sun" (see Popper 1972, 357). In a paper in 1957 Popper stated that the law (1) can be obtained as a good approximation by making false assumptions. Newton's theory unifies Galileo's and Kepler's, but "it corrects them while explaining them" (*ibid*., 198–202).

Let M be the mass of the Sun, and m the mass of the planet at the distance r from the Sun. Using the "skinny triangle approximation" $\tan \theta \approx \theta$, one can show that for a planet moving with velocity v along a nearly circular orbit the centripetal force towards the Sun is mv^2/r. By Newton's second law (force = mass × acceleration), the acceleration of the planet towards the Sun is v^2/r. The Law of Gravitation gives the force by which the Sun attracts the planet and the force by which the planet attracts the Sun. These forces are proportional to the product Mm divided by the square r^2. Again combining with Newton's second law, the net acceleration of the planet towards the Sun equals $G(m + M)/r^2$, where G is a constant. Putting these two equations for the acceleration together, we obtain $v^2 = G(m + M)/r$. For near circular orbits the velocity v is approximately $2\pi r/t$, so that $v^2 = 4\pi^2 r^2/t^2$. Hence, $t^2/r^3 = 4\pi^2 / G(m + M)$. This simplified approximate derivation shows that Newton's theory entails the conclusion

(2) $t^2/a^3 = 4\pi^2/G(m + M)$.

This result is still idealized, as it does not take into account the perturbations by other planets. But it corrects Kepler's original empirical law (1), since the ratio t^2/a^3 depends on the mass m of the planet, and thus is different for planets with different masses. For example, the Jupiter mass is 318 times the Earth mass. Thus, if (2) is correct, then (1) is mistaken. Still, the differences of (2) for the planets in our solar system are not large, since their masses m are small in relation to the mass M of the

Sun. For example, the Solar mass is 1048 times the Jupiter mass. The law (2) entails two idealizational laws:

(3) If the masses of two planets were approximately equal, the values of t^2/a^3 for them would be approximately equal.
(4) If $m = 0$, then $t^2/a^3 = 4\pi^2/GM = \text{constant}$.

In fact, in many textbooks Kepler's original law (1), or the consequent of the counterfactual conditional (4), is derived directly from Newton's theory by assuming that the Sun is a stable massive body whose acceleration toward the planet by gravitation can be ignored. The corrected version (2) may be useful in the application of celestial mechanics to moons of planets, binary stars, and satellites.

Carl G. Hempel referred to Duhem and Popper in his remark that "while, strictly speaking, Newton's law contradicts Galileo's, it shows that the latter is almost exactly satisfied in free fall over short distances", so that "it might be said to provide an *approximative D-N explanation* of Galileo's law" (see Hempel 1965, 344; cf. Niiniluoto 1990b, 37–39). Similarly, we may say that Newton's theory *approximately explains* Kepler's Harmonic Law (1), since it deductively explains (2) and the distance between the laws (1) and (2) is small. Further, if the law (2) is closer to the truth than (1), then this approximate explanation is also a *corrective explanation*.[3]

These considerations can be generalized to many important methodological contexts (see Niiniluoto 1990b). What is needed is a criterion for saying that a scientific law approximates another law. If this criterion is based on a reflexive and symmetric distance or metric, two laws which are sufficiently close to each other are *approximate counterparts*. Then we can say that theory T *approximately explains* law L if T explains an approximate counterpart of L, i.e. T explains L´ which is close to L. Two theories T and T´ are approximate counterparts if for each law in T there is an approximate counterpart in T´. Theory T *approximately reduces* to theory T' if T reduces to an approximate counterpart of T´.

Let E be empirical data, obtained by observation, measurement, or experimentation. The distance between laws can be modified so that it applies to pairs of empirical data. Then we can say that theory T is *approximately successful* relative to empirical data E if T entails an approximate counterpart of E. If theory T is approximately successful relative to two independent empirical statement E and E´, which are either explained or predicted by T, then T achieves approximate unification (cf. Sect. 6.4). The relation of Newton's mechanics to Kepler's and Galileo's laws is an example of such approximate unification.

[3]Rott (1989) compares several ways of analyzing approximations and idealizations in the Kepler-Newton case.

8.2 Verisimilitude and Legisimilitude

Let L* be the *true* law governing some phenomenon. Such laws of nature may be qualitative or quantitative, deterministic or probabilistic, theoretical or observational. Laws are stronger than mere accidental generalizations, but philosophers do not agree how to characterize their "lawlikeness". Often laws are required to sustain counterfactual conditionals and express physically or nomically necessary connections between attributes or quantities. Thus, relative to a discrete conceptual space or a continuous state space, a law of coexistence specifies what is physically possible and what is physically necessary. A law of succession, typically a dynamic or causal law, specifies physically possible trajectories of events (cf. Niiniluoto 1987).

L. J. Cohen (1980) suggested that, instead of *verisimilitude* which may concern closeness to any kind of truth about reality, philosophers of science should be interested in *legisimilitude*, i.e. similarity of theoretical claims with the true laws of nature. A solution to this challenge can be given by choosing a true quantitative law L* as the target, and to specify distances between such laws on the basis of their similarity (see Niiniluoto 1990a,b). Then law L is *approximately true* if it is sufficiently close to the true law L*. A set of laws T is approximately true if the minimum distance from the members of T to L* is sufficiently small. T has a high degree of legisimilitude if the set as a whole is sufficiently close to L*, i.e. some members of T are very close to L* and T does not include too many bad alternatives.[4]

The project of defining the notion of *truthlikeness* or *verisimilitude* started with Karl Popper's attempt to give conditions for one theory (as a deductive closed set of statements in some language) to be closer to the complete truth (in the same language) than another theory (see Popper 1963, 1972). After the refutation of Popper's definition in 1974 by David Miller and Pavel Tichý, the project continued with the new "similarity approach" which employs the notion of resemblance or similarity between states of affairs. The truthlikeness of a first-order theory H depends on its distance from a target C*, which could be chosen as the most informative true generalization (i.e. complete theory) in the given language L.[5] In the simple special case, the monadic language includes qualitative one-place predicates, which define a classification system of mutually exclusive Q-predicates. Then the target is the true monadic constituent in L which specifies which Q-predicates (kinds of individuals) are exemplified and which are not (see (7.1)), and the Clifford distance between constituents is measured by the number of matches in their claims. Thus, if CT_i is the set of the positive existence claims of constituent C_i, then the Clifford distance between C_i and C_j is the cardinality of the symmetric difference

[4]For details, see Niiniluoto (1987), Ch. 11.

[5]Alternatively the target could be the true state description in a monadic language L, where a state description specifies which Q-predicate each individual in L satisfies. All sentences of L have a normal form as a disjunction of state descriptions. For the treatment of truthlikeness of singular statements along these lines, see Niiniluoto (1987), Ch. 8.

between CT_i and CT_j.[6] For some purposes, it is useful to modify the Clifford measure by allowing the distance between constituents to depend on the distances between Q-predicates. As Hintikka has shown, in predicate logic with relations all generalizations have a "distributive normal form" as a disjunction of constituents which tell which kinds of sequences of individuals (of a given length) can be drawn from the universe. Thus, a first-order theory H typically allows several alternative mutually exclusive constituents (complete descriptions of possible worlds). Then theory H is *approximately true*, if its minimum distance from the true constituent C* is sufficiently small. This minimum distance is zero if H is true, i.e. C* entails H. Theory H is *truthlike*, if all of the allowed alternatives are close to C*. This means that truthlike theories are also approximately true, but not conversely, Hence,

(5) Theory H is approximately true to a maximal degree if and only if H is true.
(6) Theory H is maximally truthlike with value 1 if and only if H = C*.

More precisely, when H is expressed as a disjunction of constituents $C_1, \ldots, C_m$, then its *degree of truthlikeness* Tr(H,C*) is function of the distances from $C_1, \ldots, C_m$ to C*, where $0 \leq \text{Tr}(H,C^*) \leq 1$. For Graham Oddie (1986), this function is the average distance, while for Ilkka Niiniluoto (1987) it is the weighted average of the minimum distance and the sum of all distances.[7] The main difference between these explications is that Oddie's average measure fails to satisfy, while Niiniluoto's min-sum measure satisfies, Popper's fundamental adequacy condition

(7) Among true theories, truthlikeness increases with logical strength.

Unlike Popper's comparative definition, both of these explications allow that some false hypotheses are so close to the truth that they are cognitively better than trivial truths like tautologies, which simply express our ignorance. By this feature they can be used to characterize *scientific progress* as increase of truthlikeness, where typically a false theory is replaced by an improved false theory which is closer to the truth (see Niiniluoto 1984, 2017).[8]

---

[6]This measure can be applied also to nomic monadic constituents, which claim that some Q-predicates are physically possible and some physically impossible. In the definition of legisimilitude, it is natural to take the true nomic constituent as the target. This is in harmony with the modal treatment of quantitative laws which express the physically possible combinations of quantitative properties (see Niiniluoto 1987, 95, 112). The theories of Kuipers (2000) can be interpreted as nomic constituents (see Niiniluoto 1987, 381).

[7]See also Niiniluoto (1998, 1999a) and Kuipers (2000).

[8]Rowbottom (2015) argues that scientific progress is possible in the absence of increasing verisimilitude. He asks us to imagine that the scientists in a specific area of physics have found the maximally verisimilar theory C*. Then it may seem that no more progress is possible, but yet this general true theory could be used for further predictions and applications. One reply to this argument is that predictions from C* constitute new cognitive problems for the scientists. Moreover, on the basis of conceptual pluralism, in Rowbottom's thought experiment it would still be possible for the physicists to achieve further progress by extending their conceptual framework and replacing the target C* with a stronger truth in order to find a still deeper truths about their research domain.

As by (7) truthlikeness covaries with logical strength among true theories, we have a result which is appealing to theoretical realists. Let T be a true theory in language L with theoretical terms, and let T_0 be its observational subtheory in the observational language L_0 – e.g. T_0 could be the Craigian transcription of T in L_0 (see Tuomela 1973). Then T is more truthlike than T_0.

For false theories, logical strength alone cannot decide closeness to the truth: if H entails G, then G need not be more truthlike than H. For the same reason, truthlikeness is not generally preserved under deduction: if $H \vdash G$, the weaker theory G may contain constituents which are far from the truth. But the deductive consequences of a truthlike theory have to include constituents which are close to the target. Further, minimum distance from the truth cannot decrease in deduction, so that approximate truth is preserved under deduction for statements in the same language:

(8) If H is truthlike in L, then its deductive consequences in L are approximately true.
If H is approximately true in L, then its deductive consequences in L are approximately true.

The result (8) has to be qualified, if we consider consequences from a general theory H and singular initial conditions (cf. Niiniluoto 1999a, 195). If the true theory H is a rapidly growing quantitative function, then even small errors in the initial conditions may be multiplied so that the consequences are far from the truth. This is the case in nonlinear chaotic systems (see Sect. 4.2). But as the case of coin tossing shows, even in such systems quite reliable statistical explanations and predictions are possible. Roman Frigg et al. (2013) have further argued that weather and climate models as nonlinear dynamical systems are unpredictable, if there is only the slightest model imperfection in the basic equation. Still, short-run predictions (with few iterations) are not excluded in these systems, so that the same holds for explanations. What Hempel called self-evidencing explanations seem to be possible within such idealized models (see Hempel 1965, 274–276). For example, in a single toss of a coin heads and tails are equally probable and thus unpredictable outcomes, but when heads has occurred we can infer backward to the kind of initial condition which has produced this result. More generally, if we know what state the system has reached, we may be able to reason inversely to its initial state, and then use it as the initial condition of an explanatory argument.

Let us consider in more detail distances between quantitative statements. The simplest singular quantitative statements attribute a quantitative property to an object, so that they can be represented as points in a state space Q generated by real-valued quantities. For illustration, assume that Q is a subclass of the real plane $\mathbb{R}^2$. Then the geometrical distance between two points $x = \langle x_1, x_2 \rangle$ and $y = \langle y_1, y_2 \rangle$ in Q can be defined by the *Manhattan* or *city-block metric*

(9) $d_1(x, y) = |x_1 - y_1| + |x_2 - y_2|,$

the *Euclidean metric*

(10) $d_2(x, y) = [(x_1 - y_1)^2 + (x_2 - y_2)^2]^{1/2}$

or by the *Tchebycheff metric*

(11) $d_\infty(x, y) = \max(|x_1 - y_1|, |x_2 - y_2|)$.

With each choice of d, (Q,d) is a metric space. For comparative purposes, we obtain the following standard:

(12) x´ is better approximation of y than x iff d(x´,y) < d(x,y).

If the truth is represented by the point x* in Q, then the notions of approximate truth and truthlikeness for points x in Q can be defined by the condition $d(x,x^*) \leq \varepsilon$, where ε is the threshold of approximation. This definition can be extended to distances d(H, x*) of connected subsets H of Q from the truth x* (see Niiniluoto 1987, 303–309).[9]

Consider now simple quantitative laws which express the functional dependence of a quantity y on another quantity x. Assume that the true connection between y and x is expressed by the equation $y = f^*(x)$, where $f^*: \mathbb{R} \to \mathbb{R}$ is a continuous function. The relevant rival lawlike hypotheses are also continuous functions $g: \mathbb{R} \to \mathbb{R}$. The distance between two such functions f and g can be defined by the family of *Minkowski metrics* or *L^p-metrics* for function spaces:

(13) $L^p(f, g) = [\int |f(z) - g(z)|^p \, dz]^{1/p}$.

(See Niiniluoto 1987, 385–386.) The choice $p = 1$ gives a generalization of the city-block metric (9), $p = 2$ the Euclidean metric (10), and $p = \infty$ the Tchebycheff metric (11). According to L^1, two functions are close to each other if the volume between their curves is small. L^2 requires that the average distance between the curves is small, and L^∞ that the maximum distance between the curves is small.

The values $L^p(f,g)$, for $p < \infty$, are finite, if integration in (13) is restricted to a finite interval [a,b]. For comparative purposes, we may then say that law f_1 is *L^p-closer* to

[9]Chang Liu (1999) has made the pertinent observation that this treatment of singular approximation is "flat" in the sense that it assumes a global metric in a uniform state space Q and thus does take into account the influence of true laws on the structure of Q. For example, a claim that the velocity of light slightly exceeds the value 300.000 km/s cannot be approximately true, since it is physically impossible by the theory of relativity. As a possible reply, one could argue that the state space Q should be a neutral framework for comparing various kinds of hypotheses independently of assumed background knowledge. As false hypotheses can be truthlike, why could not physically impossible claims be approximately true? Perhaps a more interesting line of thought is suggested by Liu's another remark: if e.g. a linear law L is assumed to hold in R^2, the singular truth x* is a point on L, and the hypothesis y lies outside L, then the distance of y from x* should depend also on its closeness to L. This idea would allow that physical possibilities influence the metric structure of the state space in the same way as the distribution of masses generate non-Euclidean geometries of physical space in relativity theory. In Liu's example, the circles of Euclidean metrics around the point x* would be replaced by ellipses with L as its major axis.

g than f_2 if for some $r_0 > 0$ we have $L^p(f_1,g) < L^p(f_2,g)$ for all intervals $[-r,r]$ with $r \geq r_0$. This guarantees that f_1 is L^p-closer to g than f_2 if f_1 is uniformly closer to g than f_2 in $\mathbb{R}$.

The metric $L^2(f,g)$ can be understood as a generalization of the Least Square Difference measure in statistics, when f is a function and g is a finite set of data points (see (4.2)). Let E include the data points $< x_1,y_1>, \ldots, <x_n,y_n>$. Then the L^p-distance between function f and E is defined by

$$(14) \quad [\textstyle\sum_{i=1}^{n} (y_i - f(x_i))^p]^{1/p}.$$

In this way, all L^p-metrics can be applied to measure the distance between Newton's result (2) and Kepler's Third Law (1). The maximum difference for Jupiter is very small, as the mass m of Jupiter is only about 0.0009546 times the mass M of the Sun.

When the notion of legisimilitude is defined in terms of the distance $L^p(f,f^*)$ of law f from the true law f*, a law may be close to the truth even when its values f (x) do not hit the correct value f*(x) at any particular point x. This differs from most distance measures between qualitative statements, which explicate similarity by a function of the number of matches and mismatches between such statements (cf. Cevolani et al. 2013).

Liu (1999), who speaks about "lawlikeness" in the place of our "legisimilitude", raises interesting further questions. In his example, a fluctuating law f_1 has values which are close to the values of the true linear law f*, but as a law it does not even remotely resemble f*. Another law f_2 has values at greater distances from those of f*, but its linear graph looks more like f*. By the Minkowski metrics (13), f_1 is a better approximation of f* than f_2, but for Liu this is the wrong answer. This argument suggests that approximation of laws may be related to two different questions: (i) what are the values of the true law?, and (ii) what is the correct mathematical form of the true law? An answer to (ii) may be correct (e.g. the law is a linear function), but its success for (i) may be very poor. Any continuous function can be approximated by a polynomial with arbitrarily high precision, so that a successful answer of (i) can always be found among polynomials, but as an answer to (ii) it may fail.

One way of resolving this dilemma is to acknowledge that in many methodological contexts both questions (i) and (ii) are relevant, and the overall success of an approximation should be a balanced combination of them. In Liu's example, the function f_1 satisfies (i) but fails for (ii). Note that Newton's correction to Kepler's Third Law serves both goals (i) and (ii). It increases approximate truth in the sense of Minkowski distance measures (13), and as a correction of the original empirical regularity it gives an improved mathematical formula (2) for this law.

Another method, familiar from statistical regression analysis, is to fix the mathematical form of the searched function, either by theoretical background knowledge or considerations of simplicity, and then find among this class of functions the one which gives the closest approximation to the observed data (cf. Niiniluoto 1987, 402–403). The attempt to draw a curve precisely through all observed points, which involve observational errors, would lead to "overfit", so that typically a class of relatively simple (e.g. linear or parabolic) functions is chosen, and then the "least false" among them is chosen (cf. Niiniluoto 1999a, 185). In such cases the aim is to find the best approximate explanation of the empirical data (cf. (4.4)).

Approximate true quantitative laws are able to give also approximate singular explanations of empirical phenomena.[10] Let $y = f_1(x)$ and $y = f_2(x)$ be two curves in $\mathbb{R}^2$, and assume that the data points in E lie in the true curve $y = f^*(x)$. If f_1 is uniformly closer to f^* than f_2, then $f_1(a)$ gives a more truthlike singular explanation of $f^*(a)$ than $f_2(a)$ for all points a in E. But if f_1 is closer to f^* than f_2 by the Minkowskian metric (11) for small p, then the approximate explanations by f_1 are better than those of f_2 on average.[11] The Tchebycheff metric with $p = \infty$ guarantees that the maximum error for explanations by f_1 is smaller than that for f_2.

The idea of legisimilitude can be generalized to probabilistic laws. The *directed divergence* between two discrete probability distributions p_i and q_i $(i = 1,\ldots,n)$ is defined by

(15) $$\sum_{i=1}^{n} p_i \log(p_i/q_i).$$

For two random variables **x** and **y** with density functions f_x and g_y their directed divergence is defined by

(16) $$\int_{-\infty}^{\infty} f_x(z)\log\left(f_x(z)/g_y(z)\right)\,dz.$$

(See Niiniluoto 1987, 7–8.) By choosing g_y as the true probability distribution, (16) expresses the legisimilitude of the probabilistic law f_x. For example, if the true probability of success in a single trial is p^*, then the distance of the binomial model with probability p from the truth is the directed divergence

$$p^*\log(p^*/p) + (1 - p^*)\log[(1 - p^*)/(1\text{-}p)]$$

(*ibid.*, 404). Formulas (15) and (16) can be modified so that q_i or g_y express observed relative frequencies, so that they can be used to measure the distance between a theoretical probability distribution from an empirical distribution (cf. (14)). If this distance is sufficiently small, the theoretical distribution gives an *approximate probabilistic explanation* of the empirical data. A typical example would be the use of Mendel's laws of heredity to explain an observed distribution of the offspring of different kinds of animals or plants. It is remarkable that Peirce realized the nature of approximate probabilistic explanations, as one can see from his schema (1.14) of "statistical deduction".

---

[10]Cf. the discussion of approximate predictions in Niiniluoto (1999a), 195–196.

[11]In this case, the better theory has larger *expected* predictive success, if the values of argument x are chosen randomly. Cf. the Expected Success principle in Kuipers (2000), 310.

8.3 Abduction with Truthlikeness

When Peirce introduced the notion of abduction as the inference from surprising facts to their explanation, in his formulation abduction gives a reason to suspect that a successful explanatory theory is *true* (see (1.18)). But if the best available explanation H of evidence E is approximate, as in the case of Newton's theory and Kepler's laws, it is more appropriate to tentatively conclude that T is truthlike or approximately true (see Niiniluoto 1999b; Kuipers 1999).[12] This principle might be called *inference to the best approximate explanation*:

(IBAE) If the best available explanation H of evidence E is approximate, conclude for the time being that H is truthlike.

A variant of IBAE replaces truthlikeness by the weaker notion of approximate truth:

(IBAE′) If the best available explanation H of evidence E is approximate, conclude for the time being that H is approximately true.

If degrees of truthlikeness are introduced, then the greater the fit between H and E, the larger the degree of truthlikeness of H in the conclusion of IBAE.[13]

As an illustration of IBAE, we may refer to the problem of stemmatology discussed in Sect. 4.4. In evaluating their RHM method Roos and Heikkilä (2009) compare a hypothetical stemma with the true historical lineage of texts (which is known in the case of their artificial data set). For this purpose, one needs a distance measure between bifurcating trees. When such a measure is available, the closest hypothesis to the true one is the most truthlike of the alternatives. Such distance measures are mentioned in Niiniluoto (1987), Ch. 1, but so far no one has studied examples where scientific hypotheses are structured as trees. Then the inference to the best stemma follows the schema of IBAE. In this context, the most plausible answer to van Fraassen's (1989) "bad lot argument" (i.e. the question whether our set of hypotheses includes the true one) is that the search algorithm is able to produce truthlike alternative hypotheses for consideration (cf. Sect. 7.2).

Theo Kuipers criticizes the original formulation of IBE for three reasons (see Kuipers 2000, 171). First, it does not include already falsified theories, i.e. theories incompatible with evidence. Secondly, "it couples a non-comparative conclusion, being true, to a comparative premise, being the best unfalsified theory". Thirdly, it is difficult to see how it could be justified merely in terms of the true/false –distinction.

The first and third point of Kuipers are handled by the versions of IBAE and IBAE′ which allow falsified theories and introduce a link to the notions of

[12]This reformulation allows us to re-evaluate Duhem's and Popper's criticism of induction: Newton's laws are not derived from Kepler's laws by induction but by the modified model of abduction.

[13]This gives an answer to Stanford's criticism of Jarrett Leplin's (2004) account of partial truth (see Stanford 2006, 158).

truthlikeness and approximate truth. However, the second point can still be evoked against IBAE, as its premise is comparative but its conclusion is non-comparative.

Kuipers (1999) proposes an alternative to IBE which he calls *inference to the best theory*:

(IBT) If a theory has so far proven to be the best among the available theories, then conclude for the time being that it is the closest to the truth of the available theories.

The best theory is here allowed to be inconsistent with evidence, so that IBT may involve abductive revision in Aliseda's (2006) sense. The phrase "closest to the truth" can be explicated on three levels: closest to the observational truth, referential truth, or theoretical truth. For this purpose, Kuipers uses his own theory of truth approximation. In its "naïve" version, the comparative notion of "closer to the truth" involves a strong dominance condition: the better theory should have (set-theoretically) "more" correct models and "less" incorrect models than the worse theory. In the refined version, the notion of betweenness for structures helps to make sense of the idea that a theory may be improved by replacing worse incorrect models by better incorrect models (see Kuipers 2000, 2014).

The phrase "the best theory" in turn is defined in terms of empirical success. One theory is empirically more successful than another relative to the available data if it has "more" correct consequences and "less" counterexamples than the other theory. With these definitions, Kuipers is able to prove a *Success Theorem*: if theory Y is at least as similar to the truth as theory X, then Y will always be at least as successful as X relative to correct empirical data (Kuipers 2000, 160). Thus, higher truthlikeness explains greater empirical success. This means also that in the attempt to approximate the truth it is functional to use a method which is based on a *Rule of Success*: if a theory Y has so far proven to be empirically more successful than theory X, accept the "comparative success hypothesis" that Y will remain to be more successful than X relative all future data, and eliminate X in favor of Y (*ibid*., 114). In other words, it is rational to favor a theory which has so far proven to be empirically more successful than its rivals. This gives a "straightforward justification" of IBT in terms of truth approximation, which resembles Peirce's defense of induction as self-corrective inference (*CP* 2.781).

The results of Kuipers, including his Success Theorem, depend essentially on his way of explicating truthlikeness. This approach is not without its problems (see Niiniluoto 1998). Here one can make the reservation that the strong dominance condition implies that the comparative notions of empirical success and closeness to the truth define only partial orderings, so that many interesting theories are incomparable with each other. In particular, this means that if theory Y has so far been more successful than theory X, then X can never become more successful than Y in the future – the best prospect for X is to become incomparable with Y.

If the framework of Kuipers is examined in monadic predicate logic, one can see that the condition of "relative correctness" for his Success Theorem is very strong indeed (see below). In Niiniluoto's (1987) approach with quantitative degrees of truthlikeness, problems of incomparability disappear, since all theories are comparable with respect to their closeness to the truth, but on the other hand in general higher degree of truthlikeness does not guarantee greater empirical success. This is due to the fact that – according to the Clifford measure – the empirical success of a theory is a function of the number of its true existence claims, but the distance from the truth also depends on its negative existence claims (see Niiniluoto 2005a, 268–270).

A problem for Kuipers is that in many cases there will be no single theory which is better than all the available alternatives, so that a rule like IBT is inapplicable. A reply to this point is to replace IBT with a formulation of *comparative abduction*, both in premises and conclusion:

(IBT^c) If theory Y is a better explanation of the available evidence E than theory X, then conclude for the time being that Y is more truthlike than X.

(See Kuipers 2009). This kind of principle could not be formulated for IBE, since the notion of truth does not have a comparative form "truer than".

Rules IBT and IBT^c are natural generalizations of IBAE and IBAE′, since their premise does not restrict the explanation to approximate explanation. They are applicable to cases, where the consequences of highly abstract theories are only approximately verified in experiments. Such a discrepancy between H and E may be due to the fact that observational errors make evidence statement E incorrect, so that theory H offers a corrective explanation of E, but even in such situations it may be reasonable to conclude cautiously only that the theory is truthlike. Even though Newton's theory can give a corrective explanation of Kepler's and Galileo's empirical laws, quantum theory and Einstein's relativity theory show that Newton's mechanics is at best approximately true.

These rules are also applicable to cases, where the explanatory theory includes *idealizations*, i.e. counterfactual assumptions which are known to be false. An important approach to the method of *idealization and concretization* has been developed in the "Poznan School" by Leszek Nowak (1980): hidden ideal assumptions are first made explicit as antecedents of idealizational laws and then these assumptions are eliminated or relaxed by modifying the consequent by a process of concretization.[14] This idea can be illustrated by the Boyle-Mariotte law

(17) $pV = RT$

which is at best a truthlike description of the relations of pressure p, volume V, and absolute temperature T of real gases. Therefore, it is called the "ideal gas law" which correctly describes only the ideal gas model. This law ignores the influence of intermolecular attractive forces (a) and the finite size of gas molecules (b). When these hidden idealizing assumptions are made explicit, the factual but false law (17) is transformed to an *idealizational law*

[14] Twelve volumes on idealization have appeared in the book series *Poznan Studies in the Philosophy of Science* between 1990 and 2005.

(18) If $a = 0$ and $b = 0$, then $pV = RT$.

This law is concretized by removing the assumptions step by step:

(19) If $b = 0$, then $pV = RT - a/V$
(20) $(p + a/V^2)(V - b) = RT$.

Here the new factual claim (20) is known as van der Waals's law, which still may be an idealization with respect to some further hidden factors (cf. Morrison 2000, 59–61).

Nowak himself treated (18) and (19) as material conditionals with false antecedents, which are then trivially true. The problem is that this would make these statements true whatever their consequent is, so that Nowak was compelled to postulate ad hoc a new notion of truth. But if (18) and (19) are read as counterfactual conditionals, their truth (or truthlikeness) means that the consequent is true (or truthlike) in those possible worlds which satisfy the ideal condition and are otherwise like the actual world (see Niiniluoto 1990b).[15] In such possible worlds some real factors are eliminated or neutralized by assigning them with extreme values. Now we have the important result that the concretization (20) logically entails (19) and (18), so that the so called Correspondence Principle is satisfied and concretization is a method of "potential truth approximation" (see Kuipers 2000). Note also that if $a \to 0$ and $b \to 0$, then the law (20) approaches (17) in the limit. If (20) is true, then (19) and (18) are true as well, but by (7) as weaker truths they are less truthlike than (20). The initial ideal gas law (17) is truthlike if it is sufficiently close to the true concretization (20): the law (17) is strictly speaking false everywhere, but at least for sufficiently large values of it is approximately equal to the law (20). If (20) is false but truthlike, its consequences (18) and (19) are also truthlike but to a degree less than (20).

Nowak's account covers Galileo's study of free fall with the idealization that the moving body is not affected by the resistance of air. This law can be concretized, "factualized" or "de-idealized" by adding the resistance of air as a new linear force and calculating the modification by Newton's theory. Such *Galilean idealizations* are methodological in the sense that it is easier to study first a simpler framework which can later be enriched by new factors.

Nowak's (1980) treatment captures also another idea: to study free fall without other disturbances allows us to see this phenomenon in its "pure" or "essential" form. He assumes that for a given problem one can hierarchically list the primary and secondary factors which influence the variable under investigation, and idealization starts from the introduction of the essential or privileged ones and then adds step by step the initially eliminated secondary ones. This has been called the method of isolation or minimalist idealization.

It may happen that from a theory with hidden idealizations one can derive consequences which are close to observed regularities. For example, Newton's

[15] Independently of my proposal, originally presented in 1983, Michael Shaffer has treated idealization statements as counterfactuals. See Shaffer (2007).

mechanics gives an approximately correct description of middle-sized bodies moving with finite velocities. Sometimes such consequences give corrective approximate explanations of empirical laws. For example, the correction of Kepler's Harmonic Law by Newton's theory is a concretization in Nowak's sense, as the relevant factor m, which is missing in Kepler's formula (1), is introduced to the derived law (2). On the other hand, the consequences E′ of an idealized theory H may be too distant from the explanans E, so that H fails even to approximately explain E. In this sense, there is no general answer to the question whether idealized theories can serve as potential explanations: whether a theory satisfies the condition of approximate explanation depends on the case and the chosen standard of approximation.[16]

Nowak's own proposal (as an alternative to Hempel's DN model) is to narrow the gap between an idealized theory H and empirical reality E by relaxing the hidden idealizing assumptions of H: to explain E by H one has to concretize H into a new theory H′ and then derive E from H′. If ⍰ is used as the sign for concretization, Nowak replaces the initial theory H with a sequence of concretizations H, H_0 ⍰ H_1 ⍰ ... ⍰ H_k so that the Correspondence Principle is satisfied, and then explains E by combining initial conditions C with the least idealized theory H_k among the known concretization of H. If the process is successful, we reach at least an approximate explanation.

This treatment of explanation by idealized theories can be combined with the modified principles of abduction IBT and IBT^c.[17] From successful idealized explanation one may conclude that the theory is truthlike or more truthlike than its less successful rivals. Thus, an application of IBT^c gives the following principle for successful concretization:

(21) If H′ is a concretization of H, and H′ gives a better explanation of data E than H, then conclude for the time being that H′ is more truthlike than H.

(See Kuipers 1999). Similarly, for Xavier de Donato (2007) to abduce the best explanation means "to find the best concretized form of a certain law, which proved to be only approximately true".

The realist account of idealized explanation replaces Hempel's requirement that the explanans is true with the relaxed condition that the explanadum should be truthlike. This view has been challenged in recent discussions about "non-Galilean

[16]Idealized theories are sometimes called "models", so that one may consider explanation by idealized models. (Magnani and Bertolotti (2017) includes several survey articles on models and modeling.) Such models may be either sets of mathematical equations or structures which are interpretations of theoretical statements. In the latter sense, models may be compared to real systems by similarity metrics which are special cases of truthlikeness measures. For example, Weisberg (2013) uses the same method of "feature matching" as Cevolaniet al. (2013). Cf. also Niiniluoto (1987), 335–338.

[17]According to McMullin (1985), 264, the strongest grounds for scientific realism are provided by "fertile theories", where a theory T is fertile if its "de-idealization" T′ can accommodate anomalous evidence E. See also AR in Sect. 8.5.

explanations".[18] Richard Batterman and Collin Rice (2014) argue that "minimal model explanations" are able to explain the macrobehavior of the target system if they belong to same "universality class", even when their microdetails do not share any common features. Having the same macrofeatures of a universality class allows explanation even when the model does not represent any relevant microscale features of the system. This amounts to the instrumentalist idea that minimal models only save the phenomena without any requirement of their truthlikeness, so that successful "explanation" in this sense does not satisfy any principle like IBT.

Many philosophers have drawn attention to the peculiar nature of nonlinear systems. Batterman (2002) analyzes critical phenomena (such as boiling water or drop formation in liquids) which involve physical discontinuities and singular limits, where the idealized thermodynamical limit (i.e. the assumption that the number of particles is infinite) cannot be de-idealized or concretized. The reason is that without the infinity assumption no mathematical solutions of dynamic equations are possible. Still, Andrew Wayne (2011) argues that not only "Galilean, harmless, controllable" idealizations can explain, but we should allow that also non-Galilean idealizations, which are not true or approximately true of the physical system, have explanatory capacities. But, he adds, this would require that we give up the assumption that explanation presupposes successful representation. However, instead of giving up the idea of an abductive link between explanatory success and truthlikeness, the realist may grant that in some cases idealized zero or infinity assumptions may be introduced in order to make the equations solvable. Then we deliberatively sacrifice something from the truthlikeness of a theory in order to make it manageable, since otherwise no explanations or predictions could be derived from the theory.

8.4 From Empirical Success to Truthlikeness

In his well-known "confutation of scientific realism" Larry Laudan (1984) demanded the realists to show that there is an "upward path" from the empirical success of science to the approximate truth of theories – and then a "downward path" from the from approximate truth to empirical success.[19] A reply to Laudan's challenge is given in Niiniluoto (1984), Chap. 7, by appealing to a measure of truthlikeness and a method of its estimation (see Niiniluoto 1987). Here we have already seen examples of downward inferences like approximate explanations by approximately true laws (Sect. 8.1) and idealized theories (Sect. 8.3).[20] We have also illustrated the upward inference by rules like IBAE and IBT.

[18]See the survey in Niiniluoto (2018).

[19]The terms "upward" and "downward" correspond to resolution and composition in the regressive method (see Fig. 2.1 in Chap. 2), so that are also historically related to abduction and explanation.

[20]In Sect. 9.2 we shall return to a specific form of downward inference, the no miracle argument for the success of science.

We have also outlined Kuipers' (2000) reply to Laudan by his downward Success Theorem and upward Rule of Success.

But we still need to ask in what sense, in which way, to what extent, and under what conditions abduction might be justified. Let us recall that a rule like IBE is clearly intended to be a fallible rule of inference. It is too much to ask for a complete justification which would turn the ampliative IBE to a necessarily truth-preserving argument. The same holds for formulations of IBAE and IBT. In Chaps 6 and 7 we gave a *probabilistic justification* for abduction by showing that successful explanation of surprising facts confirms the hypothesis, and in many cases the posterior probability is high enough (in comparison to its rivals) to make the hypothesis tentatively acceptable. But this Bayesian treatment breaks down in cases, where the explanatory hypothesis has zero prior probability (van Fraassen 1989) or the evidence is in conflict with the hypothesis. If $P(H) = 0$, then $P(H/E) = 0$ for any evidential statement E, so that E fails to confirm H. Thus, if H is an idealization which is known to be false, its prior and posterior probabilities are zero. In the problem of curve-fitting the original observational data E is usually incompatible with the considered quantitative hypotheses H, so that $P(E/H) = 0$.[21]

Therefore, the probabilistic accounts of IBE given by the basic results (6.5), (6.6) and (6.8) are not directly applicable to rules like IBAE. These results establish *a probabilistic link between explanatory power and truth*: posterior probability P(H/E) is the rational degree of belief in the truth of H on the basis of E, and thereby confirmation (as increase of probability by new evidence) means that we rationally become more certain of the truth of H than before. But a rule of the form IBAE needs *a link between approximate explanation and truthlikeness*. New tools are needed to make sense of this link, since the approximate explanation of E by H allows for H to be inconsistent with E, so that P(E/H) and P(H/E) are zero. Similarly, a direct analysis of IBT and IBT^c in terms of probability and confirmation is excluded.

One important new tool is to define the notion of *probable approximate truth* for singular and general quantitative statements (see Niiniluoto 1987, 280). A theory is approximately true if it allows states of affairs that are close to the truth. If we fix a small distance δ to indicate what is to be counted as sufficient closeness to the truth, let H^δ, the "blurred" version of H, be the disjunction of hypothesis H with states of affairs that are at most within the distance δ from H. Thus, H is approximately true if and only if H^δ is true. In particular, a true theory is always approximately true. Hence, the probability PA(H) that a theory H is approximately true (within the specified degree δ) can be defined by the probability of the blurred version of H:

(22) $PA(H) = P(H^\delta)$.

[21]The standard Bayesian trick to allow that $P(E/H) > 0$ is to add to regression models H a random error with a Gaussian probability distribution (cf. (4.5)). See Bandyopadhyay and Boik (1999).

Hence, PA(H) is always greater than P(H), but it is important that the probable approximate truth PA(H) of H may be greater than zero even though $P(H) = 0$.[22] This helps us to give a reply to van Fraassen (1989), who argues that the result (6.5) about abductive confirmation does not apply to hypotheses H with a zero prior probability (see Niiniluoto 1999a, 188). Note first that Bayesian statistics treats hypotheses with zero probability by introducing a prior probability distribution on the set of hypotheses **H** and applying Bayes's Theorem for distributions. To accept a *sharp* null hypothesis H then is often taken to mean that we behave (make decisions) *as if* H were true. However, the notion of probable approximate truth allows us to assess the cognitive virtue of H in spite of its zero probability.[23] For example, if y is the mean of n measurements when the observational error satisfies the Gaussian normal distribution around the true parameter θ, one can show that the probable approximate truth of the hypothesis $y = \theta$ approaches to one, when the sample size n grows without limit (see Niiniluoto 1987, 283).

Unlike truthlikeness, which increases with logical strength among true theories (see (7)), probability P and probable approximate truth PA always decrease with logical strength:

(23) If $H \vdash G$, then $P(H) \leq P(G)$ and $PA(H) \leq PA(G)$.

The notion of high probable approximate truth satisfies the principle of Special Consequence SC: if PA(H/E) is high and H entails H´, then PA(H´/E) is high. The notion of confirmation as positive probabilistic relevance can be modified by replacing P with PA: let us say that E *ap-confirms* H if $PA(H/E) > PA(H)$ (see Festa 1999). Then (6.5) can be generalized to the following result:

(24) If $PA(H) > 0$, and H^{δ} logically entails E, then E ap-confirms H.

(See Niiniluoto 1999a, 188.) Note, however, that this kind of result does not yet justify IBAE´, since in this case H is compatible with E.

Another tool in connecting objective degrees of truthlikeness and epistemic matters is based on the idea of estimating verisimilitude by the expected degree of truthlikeness, which is an attempt to solve the epistemological problem of verisimilitude (see Niiniluoto 1987, 269; cf. Popper 1972, 58, 335).[24] If $C_1, \ldots, C_n$ are the alternative constituents (complete states of affairs) expressible in some language, and the degree of truthlikeness of theory H would be $Tr(H,C_i)$ if C_i were the true state, then the *expected verisimilitude* ver(H/E) of H given evidence E is defined by

[22] The function PA can be used to explicate the notion of "probable approximate correct" or PAC-learning in machine learning (see Niiniluoto 2005b).

[23] Our reply is thus different from Abner Shimony's (1970) "tempered personalism", which argues that all "seriously proposed" hypotheses should have non-zero probabilities.

[24] Zamora Bonilla's (1996) alternative solution is to measure directly the distance between a theory and empirical laws, so that for such evidence-relative notion of truthlikenesss he does not need the objective notions of truth and truthlikeness at all. Schippers (2015) develops a "two-sided" coherence measure (cf. Sect. 6.4) for the distance between theory and evidence.

(25) $\text{ver}(H/E) = \sum P(C_i/E)\,\text{Tr}(H, C_i)$,

where the sum goes over all i = 1, ..., n. If all constituents C_i are equally probable given E, then also their ver-values are equal. But if evidence E entails one of them, say C_1, then ver(H/E) = Tr(H,C_1). By the same token, if $P(C_1/E) \to 1$ with increasing evidence, then ver(H/E) → Tr(H,C_1). In Hintikka's inductive logic, all probability concentrates asymptotically on the constituent C^c which states that the universe is of the same kind as the sample E of size n with c exemplified Q-predicates (see Sect. 8.1), so that

(26) $\text{ver}(H/E) \to \text{Tr}(H,C^c)$, when $n \to \infty$ and c is fixed

(27) $\text{ver}(H/E) \to 1$ if and only if $H = C^c$, when $n \to \infty$ and c is fixed.

The value of ver(H/E) generally differs from the probability P(H/E). For example, the ver-value of a tautology is less than maximal even though the probability of a tautology is 1. This measure has also the important feature that ver(H/E) may be non-zero, and even high, even though E refutes H and P(H/E) = 0.[25] For example, by (27) ver(H/E) is asymptotically close to one for constituents which are close to C^c.

Function ver is a generalization of the notion of epistemic posterior probability in the following sense: ver/H/E) is the expected degree of truthlikeness of H given E, while P(H/E) is the expected truth value tv(H) of H given E. Namely, if tv(H,C_i) is the truth value (1 for truth, 0 for falsity) of H if C_i were true, then

$$P(H/E) = \sum P(C_i/E)\text{tv}(H, C_i).$$

The definition (25) can be relativized to background knowledge B, so that the value ver(H/E&B) is calculated with conditional probabilities P(C_i/E&B). This means that the sum in (25) includes only those constituents C_i which are compatible with B.

We may generalize the notion of probabilistic confirmation in two ways. The counterpart of the High Probability (HP) criterion of confirmation is *High Expected Verisimilitude* (HEV):

(HEV) Evidence E HEV-confirms theory H if ver(H/E) is sufficiently high.

The counterpart of the Positive Relevance (PR) criterion is *Increase of Expected Verisimilitude* (IEV):

(IEV) Evidence E IEV-confirms theory H (relative to background assumptions B) if ver(H/E&B) > ver(H/B)

(see Festa 1999). Then, for example, if H entails E but ¬H does not entail E, then E IEV-confirms H. This implies that E IEV-confirms H if H is the only deductive explanation of E (see Niiniluoto 1999a, 186). It is important to note that, unlike

[25] Popper's proposal of interpreting his degrees of corroboration - variants of the relevance measures of confirmation (see Sect. 6.2) – as indicators of verisimilitude (see Popper 1972, 103) fails precisely on this point, since such degrees have the value zero as soon as the hypothesis is falsified by evidence.

probabilistic notions P and PA (see (23)), the function ver of expected verisimilitude (and the related notions of confirmation) may favor the stronger one of alternative hypotheses. This is dramatically seen in the result (27), where ver(H/E) is maximal for the logically strongest hypothesis C^c compatible with evidence E.

The functions PA and ver serve as tools for analyzing confirmation in the context of approximate reasoning. Suppose that H approximately entails G by implying an approximate counterpart G′ of G, where G and G′ and incompatible.[26] Then P(G′/H) = 1 but P(G/H) = 0. In this case we have P(H/E) ≤ P(G′/E) ≤ PA(G/E) for any evidence E, which can be expressed as a modification of the principle of Special Consequence SC of Sect. 6.2:

(28) If H approximately entails G, and E HP-confirms H, then PA(G/E) is high, i.e. G is probably approximately true given E.

Further, if H approximately entails E, then the blurred version E^δ of E is positively relevant to H:

(29) If H approximately entails E, then E^δ PR-confirms H.

The conclusion of (29) implies that H PR-confirms E^δ, so that H ap-confirms E. This result is general in the sense that H may be a quantitative law or theory. The notion of HEV-confirmation does not satisfy Special Consequence SC, but a modification of SC for the ver-function is the following:

(30) If ver(H/E) is high, and H′ is an approximate counterpart of H, then ver(H′/E) is relatively high.

With the ver-function we can also formulate a rule of acceptance:

(31) Given evidence E, accept the hypothesis H with maximal value ver(H/E).

In the spirit of cognitive decision theory, maximizing expected verisimilitude is a direct generalization of the principle of maximizing posterior probability.[27] In fact, Niiniluoto's min-sum-measure of truthlikeness is a generalization of Levi's (1967) assignment of epistemic utility: for a disjunctive hypothesis, truth value is replaced by one minus the minimum distance from the truth, and information content by one minus the normalized sum of the distances of disjuncts from the truth. If all basic alternatives are equally distant from the truth, so that the distance function is trivial, the min-sum measure reduces to Levi's proposal (Niiniluoto 1987, 417–419). Note that P(H/E) would be trivially maximized by choosing H as E, but this is not generally true for ver(H/E), since ver(H/H) < 1 when H does not have maximal content, i.e. H is equivalent to a disjunction of at least two constituents.

[26]Such approximate steps may occur in inter-theoretic reduction, so that their probabilistic analysis should be complemented by the notion of PA. See e.g. the notion of "analogy" in Dizadji-Bahmani et al. (2011).

[27]The Bayesian decision method of point and interval estimation by minimizing expected loss can be reinterpreted in terms of maximizing expected truthlikeness, but then the sum (25) has to be replaced by an integral (see Niiniluoto 1987, 426–441).

In order to reply to Laudan's challenge concerning the upward path, we should investigate the conditions under which the following kinds of principles are valid:

(32) If H′ is a better approximate explanation of E than H, then ver(H′/E) > ver (H/E).
(33) If H approximately explains E, and H may incompatible with E, then the expected verisimilitude of H given E is high.
(34) If H approximately explains E, then E IEV-confirms H.

If these principles hold, at least on some additional conditions, then (relative) explanatory success gives us a rational warrant for claims about (relative) truthlikeness. Note that (34) is generalization of the Basic Theorem (6.6) for explanation and probabilistic PR-confirmation.

For quantitative laws and theories similar principles can be approached by results like (28) and (29) which show that successful approximate explanation increases the probability of a theory – and thereby increases the probable approximate truth of its approximate counterparts. The principles (32)–(34) can also be studied and defended in the context of inductive logic, where the evidence E is expressed in the same monadic language L as the relevant hypotheses, i.e. constituents C_i of L (see Niiniluoto 2005a, 271). When the sample size n is sufficiently large, ver(C_i/E) decreases with the Clifford distance between C_i and C^c (see Niiniluoto 1987, 342), so that

(35) If C_i is closer to C^c than C_j, then, for large values of n, ver(C_i/E) > ver(C_j/E).

Thus, if C_i approximately explains E, then C_i entails an empirical statement E′ which differs from E only in a small number of Q-predicates. By (35), C_i is a better approximate explanation of E than C_j if and only if C_i is closer to C^c than C_j, so that we have a proof of the comparative principle (32). By the convergence result (26) ver(C_i/E) is approximately equal to Tr(C_i,C^c) for large evidence E. If now C_i approximately explains E in the sense that the distance between C_i and C^c is small, then ver(C_i/E) is relatively high, and thereby we have a proof of the result (33). By the same reasoning, E increases the ver-value of H in this case, so that we have a proof of the result (34). These results can be generalized to cases where the rival explanations are the universal generalizations of L.[28] Again, asymptotically the best explanation of E is the boldest generalization C^c compatible with E, but ver(H/E) can be relatively high when H is sufficiently close to C^c (see Niiniluoto 1987, 341–345). It is interesting to note that both the symmetric (32) and non-symmetric principle (33) can be justified in the same framework.

If all the relevant hypotheses are idealizations (i.e. defined relative to some counterfactual assumption B), then ver(H/E) has to be calculated indirectly. One way of doing so is to replace E with E(B), which is what E would have been if B

[28]Similar arguments can be applied to existential-universal generalizations which make some positive claims and some negative claims about the Q-predicates, but may leave some Q-predicates with question marks (see Niiniluoto 1987, 337; 2005a, 269). Such statements, which include constituents as a special case, are called c-theories by Cevolaniet al. (2013).

were true. With this modification, principles (32) and (33) can be applied to situations involving idealized hypotheses (see Niiniluoto 1987, 285–288).

Laudan (1984), 119, wonders whether my approach to his challenge involves a confusion between "true verisimilitude" and "estimated verisimilitude". I plead not guilty, since this is a distinction which I have stressed and explicated ever since I started to work with this concept in 1975. Laudan is right in stating that "estimated truthlikeness and genuine verisimilitude are not necessarily related", but this is what any fallibilist should expect. The abductive upward path is ampliative, so that its premises are not "necessarily related". What the function ver hopefully does in a principle like (33) is to establish a rational but *fallible bridge between the approximate explanatory success of H and the truthlikeness of H*. This is a generalization of the probabilistic bridge established between explanation and truth in the Basic Theorem (6.5).

Another way of replying to Laudan's doubt that estimated verisimilitude is "accessible but nonepistemic" is to emphasize that definition (25) gives a rational estimate of the real truthlikeness of H, and this estimate is as "epistemic" as posterior probability P(H/E) in the estimation of truth value. Laudan expressed his doubt before I had pointed out that function ver estimates truthlikeness also in the following strong sense: if evidence E is true and in the limit gives exhaustive information about the alternatives C_1, ..., C_n, then ver(H/E) approaches the actual degree of truthlikeness Tr(H,C*) of H relative to the true state C* (see Niiniluoto 1987, 276). In other words, given a condition of *evidential success* (ES) to the effect that evidence E is true and fully informative about the variety of the world, the convergence results (26) and (27) in inductive logic can be reformulated in terms of *convergence to the truth*:

(36) Given that ES holds, ver(H/E) $\rightarrow$ Tr(H,C*), when $n \rightarrow \infty$ and c is fixed
(37) Given that ES holds, ver(H/E) $\rightarrow$ 1 if and only if H = C*, when $n \rightarrow \infty$ and c is fixed.

These results illustrate the Peircean insight that our interaction with the world by means of observation and experimentation has crucial significance for the defence of realism (cf. Niiniluoto 1999a, 94). The chance that ES is correct can be improved by experiments, which by manipulating nature can realize otherwise hidden possibilities (cf. Hacking 1983). But by the nature of ampliative inference we can never have conclusive logical reasons for excluding the possibility that ES is incorrect in a particular situation. In this sense, (37) gives the strongest possible convergence result in the most favorable epistemic situation.[29]

The situation is more complicated if the hypotheses are expressed in a language L′ with theoretical predicates, while evidence E is expressed in the observational language L. Then there will be several theoretical constituents (and several universal

[29]The convergence results (36) and (37) can be relativized to a true background theory B, so that ver (H/E&B) converges to the truth given ES and B, but as fallibilists we cannot be completely certain that B is true.

generalizations) in L´ which entail C^c in L and therefore asymptotically receive non-zero posterior probabilities given E. But even with such "incomplete evidence" some theoretical generalizations may have relatively high degrees of estimated verisimilitude (Niiniluoto 1987, 275, 345). Let us illustrate these issues in the case, where L´ includes one additional monadic theoretical predicate M (see Niiniluoto 2005a, 270; cf. Niiniluoto and Tuomela 1973). Then each Q-predicate Q_j of the observational language L is split into two Q-predicates of L:

$$Q'_{j1}(x) = Q_j(x)\&M(x)$$
$$Q'_{j2}(x) = Q_j(x)\&\neg M(x).$$

Then, by (7), the true constituent of L´ is more truthlike than the true constituent C* of L. But there are also several false constituents of L´ which entail C*. For example, assuming that in fact $Q'_{j1}(x)$ is non-empty and $Q'_{j2}(x)$ is empty, a false constituent may mistakenly claim that $Q'_{j1}(x)$ is empty and $Q'_{j2}(x)$ non-empty, but still correctly entail that $Q_j(x)$ is non-empty. Kuipers (2000), 213, excludes such constituents with his assumption of "relative correctness", which seems to be ad hoc as a general principle. But a local assumption B, which entails $(\forall x)(Q_j(x) \rightarrow M(x))$, may be justified in some theoretical contexts.[30] With this kind of background assumption B about the relations of L and L´, i.e. by using the values ver(H/E&B) for hypotheses H in L´, we can go through the arguments in favor of the principles (32), (33), and (34). This example indicates more generally how "bridge principles" like B help to establish fallible links between empirical evidence and claims about the verisimilitude of theoretical claims.

8.5 Abductive Belief Revision

In this section we combine the work on abduction and truthlikeness with the logical program of studying belief revision (BR). The best known BR approach is the AGM-theory, developed by Carlos Alchourrón, Peter Gärderfors, and David Makinson (see Gärdenfors 1988; Hansson 1999). In BR models, belief sets are deductively closed sets of sentences (usually in propositional logic), and changes of beliefs are induced by input sentences. If K is a belief set, then Cn(K) = K, where Cn is the operator of logical consequence. If input A is consistent with K, then A is added to K. This procedure defines the *expansion* K + A = Cn(K U {A}) of K by A. If input A contradicts K, i.e. K contains ¬A, then the *contraction* K – A of K with respect to A is obtained by retracting from K the sentence ¬A and other sentences which logically entail ¬A. The *revision* K*A of K by A is defined by the Levi

[30]Connections between theoretical and observational predicates have been called "reduction sentences" (Carnap), "rules of correspondence" (Nagel), and "bridge principles" (Hempel) (see Niiniluoto 1999a, 112). For example, in Wilson's cloud chamber a curved trail of ionized gas particles is a manifestation of the negative electric charge of an electron.

identity $K^*A = (K - \neg A) + A$. It follows that in this model the new information A is always trusted, so that A belongs to K*A. Only in models of "non-prioritized belief revision" this success postulate is not generally valid.

If (finitely axiomatizable) theories are expressed as sentences in a first-order language, then the expansion, revision, and contraction of theory T by input A can be defined by

(38) $T + A = T \,\&\, A$
$T^*A = \vee D_T(A)$
$T - A = T \vee T^* \neg A,$

where $D_T(A)$ is the set of constituents in A which are closest to T (see Niiniluoto 2011b).[31]

The guiding principles of the AGM-theory are consistency and informational conservatism: in modifying your beliefs in the face of new information you should secure consistency while maintaining as many of your old beliefs as possible. The notions of truth and falsity are taken to be irrelevant or inaccessible (see Gärdenfors 1988, 20). As a model of scientific change, BR's motivation is coherentist epistemology or instrumentalist philosophy of science.[32]

If belief revision is viewed from the perspective of truth approximation, the first surprising observation is that the expansion of a false theory T by true input A need not increase the truthlikeness of T.[33] To see this, let T state that the number of planets is 9 or 20, where the true number is 8. Let A state that this number is 8 or 20. Then T + A is the bad theory that there are 20 planets. The same holds of the revision of false T by true B. The only safe case is the expansion of true beliefs by true inputs, since then by (7) T + A is more truthlike than T, but this case of cumulative growth is not very interesting or realistic as a model of theory change in science. However, in the special case of convex or "conjunctive theories", expansion and revision of false beliefs by true input information will increase truthlikeness.[34]

Aliseda's (2006) important contribution is to link BR with abduction (cf. Sect. 3.3). She formulates *abductive expansion* by the rule:

(AE) Given a background theory T and a novel observation E (where T does not entail E), construct an explanation H such that T and H together entail E, and add E and H to T.

---

[31]This definition of contraction uses an idea proposed by Adam Grove.

[32]However, Isaac Levi's work on belief revision is based on his treatment of truth and information as epistemic utilities (see Sect. 7.1), but Levi explicitly denies the relevance of truthlikeness considerations.

[33]This result is due to Niiniluoto in 1999a, b. See Niiniluoto (2011b).

[34]See papers in the special issue "Belief Revision Aiming at Truth Approximation" of *Erkenntnis* 75:3 (2011). See also Cevolani et al. (2013).

Abductive revision applies to cases where theory T and evidence E are in conflict:

(AR) Given a background theory T and an anomaly E (where T entails ¬E), revise the theory T into T′ such that T′&E is consistent, and construct an explanation H such that T′ and H together entail E, and add H and E to T′.

Gustavo Cevolani (2013) modifies these principles by allowing that a successful explanation H is not found. Thus, the result of abductive expansion is.

(AE′) T + H, if there is an abductive hypothesis H for E,
T, otherwise.

The result of abductive revision is.

(AR′) T*E + H, if there is an abductive hypothesis H for E given T*E,
T, otherwise.

These definitions allow that E does not belong to the expansion or revision of T, if the search for an explanation H of E fails, so that they belong to non-prioritized BR.

The rule of abductive revision AR can be elaborated by taking into account the so called Duhem – Quine problem (cf. Duhem 1954): a theory T entails observational predictions only in conjunction with initial conditions I, so that an anomaly E (where T & I entails ¬E) can be solved either by revising T or by revising I. This choice can be illustrated by the contemporary debate about dark matter. Already in 1933 Fritz Zwicky estimated that there is not enough ordinary matter to hold galactic clusters together, and postulated that there must additional "dark matter". Further evidence was given in 1980 by Vera Rubin, who noted that the rotation curves for the velocities of stars and gas clouds are flat instead of decreasing with distance from the galactic center. Here theory T is Newton's mechanics, and the initial condition I states the amount of ordinary baryonic matter in the universe. Anomalous evidence E, including observations about the expanding universe and the anisotropies in the cosmic microwave background radiation, has led to the explanatory hypothesis that the universe consists only about 5% of ordinary matter and the rest is dark matter and dark energy. Here theory T is kept constant, but the initial condition I is revised into a hypothesis I*E about the amount of dark matter, so that T and I*E entail E. The majority of physicists accept this abductive inference to the best explanation, but experimental attempts to verify or test the existence of dark matter have failed. Therefore, it has been claimed to be an ad hoc hypothesis (see Merritt 2017). The alternative strategy is to accept the initial condition I about ordinary matter, but to revise Newton's theory T*E, so that T*E and I entail E. Such revisions of the standard cosmological Lambda-GMD model have been proposed by modifying the Newtonian dynamics (Mordechai Milgrom's MOND in 1983) or the relativistic dynamics (Jacob Belenstein's TeVeS in 2004) for objects with very low acceleration.

The schemas AE and AR do not guarantee that the conclusion is true or truthlike. Practical rules of belief revision cannot assume that the truth values or degrees of truthlikeness of interesting hypotheses are known. The best we can do as fallibilists is to rely on expected truthlikeness, defined by formula (25), and the rule (31) of maximizing expected verisimilitude. Suppose that we have accepted theory T on the basis of evidence E, so that ver(T/E) is maximal. With new evidence A as input, we may calculate the value ver(T/E&A), which may be high when A is compatible or incompatible with T. Then the revision of T by A is the theory which is ver-maximal on E&A (see Niiniluoto 2011b).

Simon D'Alfonso (2011) suggests that expected verisimilitude can be used as a criterion of accepting inputs in non-prioritized belief revision. He applies Makinson's model of "screened revision", where the belief set K contains as its subset a core C which is immune to revision. Then an input A is accepted, if it does not conflict with C and ver(K*A/C) > ver(K/C).

Cevolani (2013) applies the ver-measure to abductive expansion and revision in the following way. The result of *verisimilitude-guided AE* is

(41) T + H, if there is an abductive hypothesis H for E such that ver(T + H/T) > ver (T/T),
T, otherwise.

The result of *verisimilitude-guided AR* is.

(42) T*E + H, if there is an abductive hypothesis H for E given T*E such that ver (T*E + H/T) > ver(T/T),
T, otherwise.

Note that by (42) abductive revision does not satisfy the Levi identity (see Schurz 2011b). Rules (41) and (42) guarantee that estimated truthlikeness is increased in verisimilitude-guided abductive expansion and revision.

Let us still illustrate abductive revision by examples with idealized theories (cf. Niiniluoto 2005a). Three different approaches to the truthlikeness of such theories are proposed in Niiniluoto (1987), 285–288. Assume now that our rival hypotheses in L are all defined relative to a presupposition B which is false (even known to be false). For example, B may be a counterfactual idealizing assumption. The rival hypotheses are defined so that their disjunction follows from B, but they are all false. In this case, the role of the complete truth C* as the target (cf. Sect. 8.2) is played by the most informative statement C*[B] in L which would be true if B were true (*ibid.*, 259–262). Suppose that we have evidence E which describes observations made in ordinary circumstances, not under the idealizing assumption B. Then the fit between E and the alternative hypotheses is not perfect.

One way of handling this situation is to transform the idealized statements C_i[B] by concretization, i.e. by elimitating the false assumption B (cf. Sect. 8.3). Then our evidence E may be directly relevant to the concretized hypotheses via the functions P and ver. By modifying and applying the basic idea of the Converse Consequence principle CC (see Sect. 7.2), we have some reason to believe in the truthlikeness of C_i[B], if E confirms its concretization C_i'[B]:

(43) If ver(C_i'[B]/E) is maximal, conclude for the time being that C_i[B] is truthlike.

The second way follows the idea of revising an anomalous observation:

(AR´´) Given a background theory T and an anomaly E, revise evidence E to E´ such that T&E´ is consistent, and construct an explanation H such that T and H together entail E´, and add H and E´ to T.

(See Niiniluoto 2005a, 262.) Suppose that there is some way of transforming E into E´ which tells what the evidence would have been under the counterfactual assumption B (see Suppes 1962). In other words, E´ is obtained from E by "substracting" the influence of those factors that are eliminated by B. For example, E gives our observations of a projectile in the air, while E´ tells how it would behave in void without the resistance of air. Then we again apply the function ver to calculate the expected value ver(C_i[B]/E´&B). If our transformation from E to E´ is reasonably reliable, then we have reason to claim on E that C_i[B] is more truthlike than C_j[B] if ver(C_i[B]/E´&B) > ver(C_j[B]/E´&B). Applying this comparative idea to alternative hypotheses g[B] relative to B, the following explication of IBAE is obtained:

(44) If ver(g[B]/E´&B) is maximal, conclude for the time being that g[B] is truthlike.

The third approach is based on the idea that, under certain conditions, we have reason to believe that our evidence E is representative of the structure of the universe. (In Hintikka's system of inductive logic, this is the case with high probability, when the sample size is sufficiently large.) Suppose that we are able to define directly the distance D(C_i[B],E) between a hypothesis C_i[B] and evidence E. The method of least square difference in curve fitting problems is an example of such a distance measure D (cf. (14)). By our assumptions, even the shortest of these distances is larger than zero. Now C_i[B] can be claimed to be more truthlike than C_j[B] on E if D(C_i[B],E) < D(C_j[B],E). Applying this comparative idea, the following explicate of the abductive rule IBAE is obtained:

(45) If D(C_i[B],e) > 0 is minimal, conclude for the time being that C_i[B] is truthlike.

In order to generalize (45) to all hypotheses g[B], the distance function D has to be extended from constituents to their disjunctions (see Niiniluoto 1987, Ch. 6.7). Another way of generalizing (45) allows that the hypotheses g[B] are expressed in a language which contains theoretical terms and thereby is richer than the observational language of the evidence E.

Chapter 9
Abduction and Scientific Realism

Abstract Scientific realism claims that theories in natural science are attempts to describe a mind-independent reality. Such theories explain and predict observable phenomena, but their content refers to entities and processes beyond the observable domain. While naive realists assert that successful theories are literally true, critical scientific realists are fallibilists who hold that even the best theories are only truthlike or approximately true. Section 9.1 argues against Laudan's pessimistic meta-induction by analyzing scientific progress as increasing verisimilitude. In the debate of scientific realists with constructive empiricists, we reject van Fraassen's scepticist assumption that the prior probability of all theories is zero. Section 9.2 defends the no miracle argument (NMA) as an instance of abduction, which concludes that the empirical and pragmatic success of theories can be explained by the assumption that they are close to the truth or at least closer to the truth than their rivals. Alternative explanations without the notions of truth and truthlikess are assessed and found to be unsatisfactory. The conclusion is that in the defence of scientific realism abduction has a crucial role: realism is the only viable explanation of the success of science.

Scientific realism claims that theories in natural science are attempts to describe a mind-independent reality. Such theories explain and predict observable phenomena, but their content refers to entities and processes beyond the observable domain. While naive realists assert that successful theories are literally true, critical scientific realists hold that even the best theories are only truthlike or approximately true. The debate of scientific realists with constructive empiricists is discussed in Sect. 9.1. The no miracle argument (NMA) is defended in Sect. 9.2 as an instance of abduction, with the conclusion that realism is the only viable explanation of the empirical and pragmatic success of science.

I. Niiniluoto, *Truth-Seeking by Abduction*, Synthese Library 400,
https://doi.org/10.1007/978-3-319-99157-3_9

9.1 Debates on Scientific Realism

Scientific realism as a philosophical position has (i) ontological, (ii) semantical, (iii) epistemological, (iv) theoretical, and (v) methodological aspects (see Niiniluoto 1999a; Psillos 1999).[1] It holds that (i) at least part of reality is ontologically independent of human mind and culture. It takes (ii) truth to involve a non-epistemic relation between language and reality. It claims that (iii) knowledge about mind-independent (and mind-dependent) reality is possible, and that (iv) the best and deepest part of such knowledge is provided by empirically testable scientific theories. An important aim of science is (v) to find true and informative theories which postulate non-observable entities to explain observable phenomena.

The thesis (i) of *ontological realism* (OR) is compatible with many philosophical positions, like various forms of materialism and dualism. In spite of his objective idealism in metaphysics, Peirce can be regarded as an advocate of OR. The main opponents of OR are subjective idealists, solipsists, phenomenalists, and social constructivists – and some logical positivists and pragmatists who deny that OR and its negation are meaningful statements (see Niiniluoto 1999a, Ch. 2). OR has been defended by appealing to common sense (G. E. Moore) and by transcendental arguments (John Searle), but for us it is most interesting to see that the strongest arguments for OR are abductive. Against subjective idealists, who claim that the table in front of me is ontologically dependent on its being perceived, the realist can point out that the table is "invariant" with respect to my senses (seeing it, feeling it, smelling it, and knocking on it), my perceptions at different moments of time, and perceptions by different persons and even recording instruments like cameras (cf. Kaila 1979). The inference from agreeing perceptions to ordinary three-dimensional physical objects is abductive in Peirce's sense, i.e. it reasons from effects to causes by the principle of common cause (see Niiniluoto 1999a, 39; Schurz 2017, 168–169).[2] The same kinds of abduction are used in science in inferences to theoretical entities. The best explanation of an observed curved path in Wilson's cloud chamber is the existence of an electron. Strong evidence for the existence of electrons comes from our ability to use their properties in engineering applications to interfere in other parts of nature (see Hacking 1983).[3] And anyone who infers by abduction to the reality of the past, in particular to entities like dinosaurs who existed before any human minds and languages had appeared on

---

[1]For a comprehensive collection of essays on scientific realism, see Saatsi (2018).

[2]Peirce's theory of meaning, expressed by his pragmatist maxim, states that our conception of objects is wholly constituted by their effects (*CP* 5.402). Hence, like the old rule states "By their fruits ye shall know them" (*CP* 5.465), the existence of various kinds of objects and their properties can be known only abductively via their effects: a sharp knife can be used for cutting wood, a force causes changes or motion and acceleration.

[3]Hacking (1983) contrasts such success in "intervening" with success in "representation", and thereby distinguishes his and Nancy Cartwright's entity realism from realism about theories. But this distinction is problematic, since both arguments for external (ordinary and theoretical) objects and arguments for scientific theories are abductive.

Earth by evolution, is committed to admit that such entities are ontologically mind-independent (cf. Niiniluoto 2014a).

Critical scientific realism can be distinguished from metaphysical or naive forms of realism by two additional theses (Niiniluoto 1999a). First, according to the principle of *conceptual pluralism*, all inquiry is always relative to some conceptual framework (just as Kant argued in his critical philosophy), but (unlike Kant thought) such frameworks can be changed, revised, and enriched. Secondly, according to the principle of *fallibilism*, all factual human knowledge is uncertain or corrigible. Even the best results of science may be false, but still they may be probable, truthlike or approximately true (see Sect. 8.2). Critical scientific realists have argued – following Peirce and Popper, *pace* opponents like W.V.O. Quine and Larry Laudan (see Laudan 1984) – that it indeed makes sense to say that one hypothetical (even false) theory is "closer to the truth" than another theory. By the same token, it is meaningful to state that a sequence of theories "approaches to the truth", even when the final limit is not reached (cf. Niiniluoto 1984, Ch. 5; Schurz 2011a). This kind of "convergent realism" does not give up the objective notion of truth, like Tarski's semantic version of the correspondence theory of truth, but uses such objective truths as the targets in the definition of verisimilitude.

Laudan's (1984) "pessimistic meta-induction" is based on the premise that many theories in the history of science have been non-referring and false but yet to some extent empirically successful. By induction, one might infer that this is the fate of our current and future theories as well. Some philosophers have continued this argument, in many cases against formulations of Psillos (1999), by giving even more historical examples of past theories which had some empirical success, including novel successes in relation to their predecessors, but still are non-referring and false by present lights. Against the claims of "preservative" or "localized" realism (see Kitcher 1993; Psillos 1999), such successes may have been based upon theoretical postulates that are discredited today (see Lyons 2006). However, a critical realist may acknowledge that, for example, relative to the historical situation the caloric theory of heat was well supported by the available evidence. Such theories were progressive in relation to their predecessors. Still, it would be too much to expect that the next theory is true (cf. Doppelt 2014). The fact that such old theories have been replaced by better but still possibly false theories is not a "Pyrrhic victory" for scientific realism (see Stanford 2006), since it supports the realist picture of scientific progress as increasing truthlikeness. For each theory mentioned in the evidence of the pessimistic induction, we can formulate an "optimistic induction" with the comparative conclusion that the successor theory is more truthlike than its predecessor (see Kuipers 2009; Niiniluoto 2017).[4] Even though many scientific theories contain idealizations, which are known to be false, the powerful method of concretization helps to remove such assumptions and thereby lead us toward the truth (see

[4]For the example of phlogiston theory, see Niiniluoto (1999a), 191¸ and Schurz (2011a); for old quantum theory, see Kuipers (2000), 278–288. For the continuity of reference in spite of meaning-variance in theory-change, see Niiniluoto (1999a), 120–132.

Sect. 8.3). This comparative and dynamic picture of progressive science evades the pessimistic conclusion that truth is a "utopian" goal of science (see Laudan 1984), i.e. that all present and future theories are far from the truth (see Sect. 8.4).

The thesis (iv) of theoretical realism is denied by instrumentalists like Duhem (1954) who claim that scientific laws and theories are conceptual tools without truth values. Laudan's *problem-solving* account allows theories to be true or false, but he opposes theoretical realism by claiming that truth is not a part of our axiology of science. The best argument against Laudan's position is to show that his notion of "problem-solving" in fact includes explanatory and predictive problems, so that success in problem-solving at least confirms the truth of the theory (see Sect. 6.3).[5]

Duhem's later followers include also Bas van Fraassen, who likes to employ the classical slogan "to save the phenomena" (see van Fraassen 1980, Ch. 3) of the instrumentalists (cf. Duhem 1969). According to van Fraassen's *constructive empiricism*, instead of accepting theories as *true*, the proper aim of science is *empirical adequacy*, i.e. in accepting a theory H we are only committed to believing that all of its observable consequences are true. Theory H may be treated *as if* it were true (see Ladyman et al. 1997), but in fact one is only interested in what H says about the observable. Van Fraassen admits that theories have truth values, and can give pragmatic explanations of facts, but this does not indicate their truth in any way. Similarly, P. Kyle Stanford's (2006) "epistemic instrumentalism" differs from Duhem's "syntactical instrumentalism" by allowing theories to have truth values, but urges that theoretical claims are "beyond our grasp".[6] Therefore, they have to deny the validity of abductive arguments from empirical success to the truth of a theory.

Peirce defended the thesis (iv) of theoretical realism against Comte's positivism by pointing that hypothetical reasoning "infers very frequently a fact not capable of direct observation" (*CP* 2.641). His position is thus based on an abductive argument.

It has been pointed out against van Fraassen that his argument against abduction is incoherent, since he has to accept abductive inferences in so far as they are confined to the observable realm: "if one has no reason to doubt commonsense abductive reasoning, then one should have no reason to doubt abduction in science" (Psillos 1999, 211). This is denied by Ladyman et al. (1997), who urge that van Fraassen rejects abduction in all cases. If this is correct, one may wonder how van Fraassen manages in everyday life! Also recall that the notion of empirical adequacy employs the notion of truth for observational statements. The rejection of abduction with respect to the observational language seems to commit van Fraassen to a phenomenalist interpretation of that language. A strong argument in this direction can be based upon Peirce's thesis that perception is already a limiting case of

[5]Indeed, Laudan's measure of the problem-solving capacity of a theory is related to Hempel's measure of systematic power (see Niiniluoto 1999a, 166). Laudan (1990) tries to avoid this problem by appealing to what Carnap (1962) called instance-confirmation: evidence confirms the next empirical prediction from the theory.

[6]For a critique of Stanford's "neo-instrumentalism", see Psillos (2009), Ch. 4.

abduction: if we assume something like a causal theory of perception, then perceptual impressions in us are caused by external objects, and perceptual judgments thus arise by abductive inferences from effects to causes. In other words, statements about observable properties of observable objects (in what was called the "physicalist" language by logical empiricists in the 1930s), and thereby the foundation of empirical science, rests on abduction. To avoid this conclusion, van Fraassen's constructive empiricist should adopt some sort of phenomenalism.

Stathis Psillos (1996) has attacked van Fraassen by claiming that inference from empirical success to empirical adequacy involves abduction. I think this is correct in the sense that such an inference is ampliative. However, it does not seem to be a genuine case of abduction (see Niiniluoto 2004, 79), since the empirical adequacy of a theory H (i.e. the truth of *all* observational consequences of H) does not *explain* the truth of the *so far* observed consequences of H. Rather, such an inference follows the pattern of enumerative induction, which involves the idea that theories are only "as-if-confirmed" and the deductive consequences of theories are confirmed, and does not satisfy the abductive principle of Converse Entailment CE. More precisely, this inference follows a modification of the principle of Special Consequence SC in Sect. 6.1:

(SC′) If evidence E as-if-confirms theory H, and H logically entails an observational statement E′, then E confirms E′.

Our counter-arguments against van Fraassen seem to indicate that his only interesting objection to abduction concerns cases where the prior probability P (H) is zero (see Sect. 7.2). But it is utterly implausible to make this assumption about all possible hypotheses that may occur to us in our ordinary lives or in science. Hence, he seems to be a kind of selective sceptic after all. In this respect van Fraassen (1998) has further developed this position by his non-standard Bayesian theory.

If H is a scientific hypothesis, the choice $P(H) = 0$ means an a priori commitment to scepticism about H, and this commitment can be claimed to be dogmatic, since updating this probability by Bayesian conditionalization always leads to zero posterior probability. Van Fraassen has suggested that *agnosticism* or "negative suspension" concerning a hypothesis H can be represented by an interval of probabilities [0,p) which includes 0 as its left point. In other words, agnosticism is "vagueness over an interval that includes 0" (see Hajek 1998; Monton 1998). This definition has the consequence that after updating by conditionalization the posterior probability of H is still of the form [0,q), but q may be larger than p. In any case, this interval still includes the point 0, and in this sense negative suspension is a position that cannot cured by empirical learning. Van Fraassen (1998) also defends the view that such a negative suspension is an "eminently suitable epistemic attitude toward science".[7] In particular, such a probability interval should be assigned to the theoretical claim about the existence of quarks.

In my view, one may question the rationality of a probability assignment which *a priori* allows one to rule out the existence of quarks, and still does this after the explanation of other particles in terms of quark theory. Even if our prior beliefs are vague, the interval (0,p) should not include 0 as its left point. But then all the results about Bayesian learning will be valid (see (6.5)), and quark theory may be confirmed or refuted by evidence. This should be the position of a "good empiricist". What is more, such an empiricism is compatible with critical scientific realism: we have seen in Sect. 8.3 that even in cases where $P(H) = 0$ or $P(E/H) = 0$ it is possible to give an account of the cognitive significance of theory H in terms of its probable approximate truth or its estimated truthlikeness: a scientific hypothesis may be known to be false, but it may still satisfy our desiderata by being closer to the truth than its rivals. The best method for searching for such good theories is to abductively propose them as explanations of surprising facts (i.e. our earlier scientific results or conflicts) and then to assess them by empirical tests.

9.2 The No Miracle Argument

For Peirce, the success of science as a cognitive enterprise is based on its use of a critical, public, and self-corrective method. He contrasted the scientific method with three dogmatic ways of settling opinions: the method of authority, the *a priori* method, and the method of tenacity (*CP* 5.358–387). In his insistence that the scientific attitude of truth-seeking requires "identification of one's interests with those of an unlimited community" (*CP* 2.654), Peirce anticipated later social epistemology. In order to be cognitively successful, so that we ultimately shall approximate the truth, scientific inquiry has to be based upon interaction with some "external permanency" or "Real things" (*CP* 5.348). And true or approximately true scientific knowledge also serves as a guide to successful practice and as a rational basis of action (*CP* 5.170, 5.199). With these ideas, Peirce became an important inspiration for later realist explanations of the success of science.

Both Thomas Kuhn and Larry Laudan – two famous anti-realists – admitted that they cannot explain why science is an effective instrument of prediction, control, and

[7]Van Fraassen (1989), 480, defends empiricism as a stance, which avoids the "inflationary metaphysics" of theoretical realists. But in claiming that a theory is empirically adequate a constructive empiricist has to go beyond what is logically implied by the data. When van Fraassen refuses to take this risk with unobservables, he appeals to a voluntarist account of rational belief as including "anything one is not rationally compelled to disbelieve" (*ibid.*, 171–172). Thus, a realist is not irrational when she infers by IBE to the theoretical entities, but it is irrational to take IBE as a compelling rule of inference (*ibid.*, 132). However, the view of IBE as compelling or obligatory is alien to the fallibilists. Even deductive rules of inference are permissions rather than obligations, and this is the case with ampliative inferences as well.

problem solving.[8] Karl Popper, in his horror of inductivism, also argued that "no theory of knowledge should attempt to explain why we are successful in our attempts to explain things" (Popper 1972, 23; cf. Hacking 1983, 57). Since the 1950s the most popular alternative among the scientific realists (among them Jack Smart, Grower Maxwell, Hilary Putnam, and Richard Boyd) is to defend realism as the best hypothesis which explains the practical (empirical and pragmatic) success of science. According to the "ultimate argument" (cf. Musgrave 1988)[9] or the "no miracle argument for scientific realism" (NMA), the ability of scientific theories to explain surprising phenomena and to yield correct empirical predictions and effective rules of action would be a "cosmic coincidence" or a "miracle" unless they refer to real things and are true or at least approximately true or truthlike (see Putnam 1978).[10] It is clear that NMA as a defence of scientific realism is an abductive argument.[11]

A forceful treatment of the "explanationist defence of realism" was given by Richard Boyd in a series of papers (see Boyd 1984). He points out that scientists use many theory-dependent methods in deriving predictions, designing experiments and testing hypotheses. Such theory-laden methods turn out to successful in practice, which can be explained by assuming that the background theories are approximately true. Psillos (1999), 78, reconstructs Boyd's argument as follows:

> The best explanation of the instrumental reliability of scientific methodology is that: the theoretical statements which assert the specific causal connections or mechanisms by virtue of which scientific methods yield successful predictions are approximately true.

This argument covers successful empirical predictions, both familiar kinds of events (e.g. eclipses by Newton's celestial mechanics) and surprising novel kinds of events (e.g. the existence of neutrino by quantum theory). It covers also reliable rules of action which are typical in applied research: from causal laws of the form "A causes B" one can derive rules like "If you want B, you ought to produce A".[12] For example, such causal laws are experimentally tested for their truth in evidence-based medicine in order to obtain therapeutic rules for the clinical treatment of patients. Science *is* empirically and pragmatically successful in this sense. The claim of NMA is then that the approximate truth of the responsible laws and theories is the best explanation of such empirical and pragmatic success of science. The form of this argument seems to be a straightforward inference to the best explanation or IBE.

---

[8]See Kuhn (1977), 332–333; Laudan (1977), 224.

[9]The term "ultimate argument" was first used by van Fraassen (1980), 39.

[10]Psillos (1999), 72–77, starts his history of NMA with Smart, who spoke about "cosmic coincidences", and Maxwell, who argued that the prior probability of realism is larger than that of instrumentalism. An early formulation of NMA was given by Eino Kaila in his defence of "critical realism" in 1926: if the correspondence between the scientific world picture and the external world did not obtain, "it would be a 'miraculous accident' that the human world picture in fact can be used to predict observable phenomena" (see the Editor's Introduction to Kaila 2014, xv).

[11]I noted this in 1978 before reading Putnam. See Niiniluoto (1984), 51.

[12]For an analysis of applied research or "design sciences" along these lines, see Niiniluoto (1993b).

Many attempts to defend scientific realism by the general no miracle argument (e.g. by Boyd, Putnam, and Psillos) appeal to forms of abduction which conclude that successful scientific theories are approximately true. In other words, instead of the traditional IBE they involve something like the modified principle IBAE′. But these authors do not make a distinction between approximate truth and truthlikeness (see Sect. 8.2), and Psillos (1999) has expressed his scepticism about the need to formalise this notion.[13] However, the continuing debates about the proper definition of truthlikeness suggest that this notion is not "intuitively clear". Without a formal explication, we do not have a reply to Laudan's "mumbo-jumbo-argument" (see Laudan 1984, 120) – and we do not know what structural properties these notions have. For this reason, it is an advantage of critical realism to use precise concepts of approximate truth and truthlikeness in the principles IBAE and IBAE′ (see Sect. 8.3).

Let us formulate the explanationist defence of realism by imitating Peirce's canonical schema (1.18) of abduction:

(EDR) Science is empirically and pragmatically successful.
If realism is true, then the success of science would be a matter of course.
Hence, there is reason to think that realism is true.

Laudan (1984) complained that this kind of inference treats realism as if it were an empirical and scientific hypothesis. But realism is a philosophical thesis about scientific methodology and theories, so that it is not true in the same sense as particular factual statements. Further, realism cannot simply state that *all* theories are true, as this would conflict with the insight of fallibilism. Theories are successful in a variety of ways, so that besides deductive explanations and predictions we have probabilistic and approximate ones. In his attack against "convergent epistemological realism", Laudan (1984), 106–107, follows Putnam (1978), 20–21, by formulating the realist position by the claim that theories are "typically" and "at least in the mature sciences" approximately true and their terms genuinely refer (*ibid.*, 106–107). But then one needs to reformulate EDR by applying abduction in the form of IBAE or IBAE′.

The modified argument EDR can be applied to specific scientific theories. Then the explanandum is the surprising fact that a theory H has been successful in explaining known observational effects and predicting novel effects (see Musgrave 1988). For example, Newton's mechanics NM is able to give approximately correct explanations and predictions of the behavior of ordinary macroscopic physical objects, and it has led to many novel predictions in new applications. Today, after the successes of relativity theory and quantum theory, we have good reasons to think that NM is strictly speaking false. Still, it is still used every day in engineering applications. So the truth of a theory is not a necessary condition for its approximate

[13]The definition of "approximate fit" by Psillos (1999), 277, is equivalent to the characterization approximate truth in Sect. 8.1, but without specification of the similarity metrics. See also Kitcher (1993), 120–123.

empirical success. To explain the continuing pragmatic success of NM, it is therefore appropriate to appeal to its truthlikeness: NM correctly postulates the existence of (something like) gravitational forces but gives only an approximately true description of the relevant laws. In Sect. 8.4 we have defended IBAE and its variant IBT (i.e. inference to the truthlikeness of the best theory H) by general arguments without making assumptions about the nature of the explanatory hypothesis H: some of the arguments establishing fallible links between empirical success and truthlikeness were independent of the question of whether H involves theoretical terms or not.

So in a general form the *no miracle argument* looks like the following:

(NMA) Many theories in science are empirically and pragmatic successful.
The truth or truthlikeness of scientific theories is the best explanation of their empirical and pragmatic success.
Hence, conclude that such successful theories are truthlike.

The first premise about the success of science is accepted both by realist and antirealists, even though in particular cases the attribution of success to a specific theory may be non-trivial. As a whole, the argument NMA involves something like the principle IBT, and the conclusion supports the position of critical scientific realism. This no miracle argument is not philosophically an "ultimate" defence of realism, since it presupposes a minimal realist framework where it makes sense to assign truth values to scientific statements (including theoretical postulates and laws). But besides semantic realists, this framework is accepted by such methodological and epistemological antirealists who think that the truth of theories is an irrelevant (van Fraassen 1989) or "utopian" aim (Laudan 1984) which "exceeds our grasp" (Stanford 2006). If successful, the no miracle argument is also relevant to those semantic antirealists and instrumentalists whose inclination to treat theories as schemata without truth values is motivated by their belief about the inaccessibility of theoretical truth.

As Kuipers has noted, the premise of IBT, unlike its conclusion, is comparative between rival theories. Therefore, one may formulate a weaker comparative version IBT^c and use it in the formulation of *comparative* NMA:

(NMA^c) Theory H′ is empirically more successful than its rival H.
That H′ is more successful than H can be explained by the assumption that H′ is more truthlike than H.
Hence, conclude that H′ is more truthlike than H.

For example, on the whole the special theory of relativity is able to explain several phenomena more accurately than classical mechanics, so that by IBT^c it is more truthlike than Newton's theory. A comparative rule like IBT^c is cautious in the sense that in seeking the most truthlike of the available hypotheses it avoids van Fraassen's (1989) problem of a "bad lot". The comparative NMA can be used to defend the realist thesis that science progresses by increasing verisimilitude: the growing empirical and pragmatic success indicates that historical sequences of theories (e.g. from Aristotle's mechanics to Buridan, Galileo, Newton, and Einstein) have in fact approached to the truth (see Niiniluoto 2017).[14]

Psillos (1999), 79, states that NMA is not a generalization of over scientists' abductive inferences, but rather a kind of "meta-abduction" which defends the general claim that science is able to reach theoretical truth by IBE. As NMA concludes that many theories are truthlike, and these theories have typically been reached by abduction, this shows that abduction or IBE is a reliable method of inference. In this way NMA is not only a defence of scientific realism but also gives support to the reliability of IBE. Let us call this the *extended* NMA.

The main objection by the antirealists to the explanationist defence of scientific realism is that the no miracle argument is viciously circular or question-begging, since it seems to defend abductive methodology by an abductive argument. Arthur Fine (1986), 161, points out that the realist is "not free to assume the validity of a principle whose validity is itself under debate".[15] Laudan (1984), 134, adds that critics of epistemic realism from Sextus to Bellarmino and Hume have known that false theories can have true consequences, so that affirming the consequent is indeed a fallacy (cf. Sect. 2.1). The realist's attempt to use a second-order abduction to defend first-order abductions is unable to convince the opponents, since it is "a monumental case of begging the question".

It is correct that NMA and NMA^c are instances and generalizations of the abductive schemas IBT and IBT^c. But our employment of the no miracle argument is not circular, since in Chap. 8 we have given a *direct* justification for IBT in terms of the rational estimation of truthlikeness by the ver-function as a generalization of the probabilistic justification of the original IBE. A further defence of the realist approach is given by Psillos (1999), 81–90, who follows Richard Braithwaite in distinguishing "premise-circular" and "rule-circular" arguments. The former attempt to justify the truth of a claim C, but explicitly presuppose C in the premises. Clearly arguments like EDR and NMA are not premise-circular, as their conclusion about realism in general or the truthlikeness of some theory in particular are not mentioned in the premises.

But the extended NMA may rule-circular in the sense that it uses abduction while its conclusion implies that IBE is a reliable form of reasoning. Psillos continues by arguing that, assuming an externalist notion of justification which allows us to rely on reliable arguments, the extended NMA is rule-circular, but not in a vicious sense. It is as harmless as using memory in the empirical defence of the reliability of memory, or using modus ponens in the meta-language to defend modus ponens in the object language. This thesis is not unproblematic (see Douven 2011), but it is not

[14]A similar argument can be given by using the measure of expected verisimilitude for retrospective comparisons, when the evidence E is taken to include our currently accepted theory T, i.e. the truthlikeness of a past theory H is estimated by ver(H/E&T) (see Niiniluoto 1984, 171).

[15]For comments on Fine's "natural ontological attitude" (NOA), see Niiniluoto (1999a), 18–20, and Psillos (1999). As a minimalist approach to ontology and truth, NOA is hoped to give an alternative to realism and antirealism, but its core position seems to a variant of realism after all.

really needed if NMA is used in a non-extended form as an argument for scientific realism. Another approach for the justification of IBE, independently of probabilistic consideration, is the defeasibility argument of Psillos (see Sect. 7.2).

The crucial second premise of NMA expresses a reply to Laudan's challenge of showing that there is a "downward path" from approximate truth to empirical success. For true theories, scientific realists have an easy answer: if a theory is true, then all of its deductive empirical consequences (if any) are true as well. As we have seen in Chap. 8, for truthlike theories the matter is more complicated, but still closeness to theoretical truth is sufficient to guarantee at least approximate, average, probable, or expected empirical and pragmatic success (cf. Niiniluoto 1984, 179–183; Kuipers 2014). Truthlikeness and approximate truth give satisfactory and even best explanations of the success of science.

This "ultimate argument" can be further developed by showing that attempts to give antirealist explanations of the success of science on the basis of pragmatism or constructive empiricism have failed. In particular, Arthur Fine (1986) has argued that in the second premise of NMA, i.e. the "downward" explanatory schema

(1) Theory H is pragmatically successful, because H is true,

an instrumentalist or anti-realist can replace the realist notion of truth by the pragmatist notion of truth. But, as the pragmatist defines truth as pragmatic success, Fine's suggestion would turn schema (1) into a non-explanatory tautology

(2) Theory H is pragmatically successful, because H is pragmatically successful.

This refutes Fine's "Metatheorem 1" (see Niiniluoto 1990a; 1999a, 197). Psillos (1999), 92–93, makes the same critical observation. He adds that Fine's later version

(3) Theory H is pragmatically successful, because H has a disposition to be pragmatically successful

needs a grounding for such a disposition, and that can be found in the approximate truth of H.

Similarly, if truth is replaced in (1) by van Fraassen's notion of empirical adequacy (see Lyons 2003), the schema

(4) Theory H is pragmatically successful, because H is empirically adequate

again fails to be explanatory, since it would "explain" the truth of some of the empirical consequences of H by the truth of all such consequences (cf. Musgrave 1988; Stanford 2000). Note further that the replacements (2) and (4) could not be made in the refined version of NMA:

(5) Theory H is pragmatically successful, because H is truthlike.

With reference to Fine, Jarrett Leplin (1987) has coined the term *surrealism* for the view that the observable world behaves *as if* our theories were true. He argues that, as an attempted explanation of the success of science, surrealism is not explanatory at all, since it simply restates the explanandum. Musgrave (1988) raises the question whether surrealism differs in any way from constructive empiricism.

Andre Kukla's (1996) distinguishes "strong surrealism" (our theories are empirically equivalent to true theories, whose deep structure is responsible for the observable phenomena) and "weak surrealism" (i.e. the observable world behaves as if our mature theories are true). While strong surrealism faces the danger of collapsing into realism, Kukla suggests that, in spite of its "unintelligibility", weak surrealism explains the success of science. Moti Mizrahi (2012) argues that weak surrealism and realism (i.e. our mature theories are true) are equally plausible as explanations for the success of science, since they yield the same independently testable predictions, so that NMA does not give the best explanation of success. However, these arguments do not escape the problem that weak surrealism only describes the fact about science (i.e. its empirical success) which is in need of explanation. Against Mizrahi, one can also point out that we have access only to the successful observable consequences of our theories *so far*, so that merely from them and the thesis of weak surrealism one cannot derive the same predictions (including surprising novel ones) as from full theories.

P. Kyle Stanford (2000) has proposed that "an epistemic antirealist", who grants the realist the defeasibility of the correspondence conception of truth, can replace NMA with the explanatory schema

(6) Theory H is empirically successful, because H is predictively similar to the true theory H*.

Here predictive similarity means that the predictions of H are "sufficiently close to those made by the true theoretical account of the relevant domain". This proposal seems to be equivalent to what Kukla calls "strong surrealism". As Psillos (2001) observes, (6) presupposes that the successful predictions can be explained in the realist sense by the true theory H*. If applied to theories H with an unknown truth value (as an epistemic antirealist should), (6) implies that there is a relation of H to the world, and the same realist explanation of success could be applied to H. Stanford applies (6) to theories which are known to be false, but then its plausibility is parasitic on the principle (5) which involves an approximate relation between a theory and the world. Stanford's final recommendation is an inclusionary reading of (6), where the empirical success of H is explained by the disjunction "H is true or H is false and predictively similar to H*", but this fails to give an independent alternative which is better than the realist explanation (5).

Gerald Doppelt (2005) claims that scientific realism must explain the "explanatory success" of science rather than its empirical adequacy. But clearly it is too much to demand with Doppelt that the truth of a theory alone would explain its "simplicity, consilience, intuitive plausibility, and unifying power". Such epistemic utilities may very well be additional desiderata that are independent of truth. For example, a tautology is certainly true, but it need not be simple, and it does not have any explanatory power. On the other hand, truthlikeness combines the ideas of truth and information, so that it helps to establish interesting links between the realist virtues of a theory and its explanatory and unifying power.

Van Fraassen (1980), 40, and Laudan (1984) have also suggested that no explanation of the success of scientific theories is needed, since theories are selected for survival by their success. No wonder that our theories "work", since we choose those theories which survive in the "fierce competition", and we use testing procedures which select reliable theories. This evolutionary or "Darwinian" move is not convincing, however, since it is phenotypical rather than genotypic (Kitcher 1993; Psillos 1999, 96). Consider an analogy: why are our cars faster today than cars fifty years ago? It is not sufficient to say that we buy faster cars than earlier (cf. van Fraassen), or that we produce in our factories faster cars than before (cf. Laudan). The evolutionary "explanation" fails to point out any characteristic permanent feature of our best theories (such as their truthlike correspondence to reality) which accounts for their ability to yield successful explanations and predictions (see Niiniluoto 1999a, 198). It is a different matter to describe the selection processes which gives us empirically successful theories and to explain why such theories are (and continue to be) successful.

Non-scientific explanations of the success of science – e.g. appeal to miracles or God's will – are not satisfactory. Therefore, we may conclude that scientific realism is the *best* and even the *only* viable explanation of the empirical and pragmatic success of science. This strong form of IBE, i.e. IOE, justifies the abductive no miracle argument NMA, and thereby gives us the strongest defence of scientific realism.

Bibliography

Achinstein, P. (1971). *Law and explanation*. Oxford: Oxford University Press.

Aliseda, A. (1997). *Seeking explanations: Abduction in logic, philosophy of science and artificial intelligence* (ILLC dissertation series). Amsterdam: University of Amsterdam.

Aliseda, A. (2000). Abduction as epistemic change: A Peircean model in artificial intelligence. In Flach & Kakas (Eds.), *Abduction and induction: Essays on their relation and integration* (pp. 45–58). Dordrecht: Kluwer Academic Publishers.

Aliseda, A. (2003). Mathematical reasoning vs. Abductive reasoning: A structural approach. *Synthese, 134*, 25–44.

Aliseda, A. (2006). *Abductive reasoning: Logical investigations into discovery and explanation*. Dordrecht: Springer.

Aliseda, A. (2007). Abductive reasoning: Challenges ahead. *Theoria: An International Journal for Theory, History and Foundations of Science, 22*, 261–270.

Anderson, D. R. (1986). The evolution of Peirce's conception of abduction. *Transactions of the Charles S. Peirce Society, 22*, 145–164.

Aristotle. (1989). *Prior analytics* (R. Smith, Trans.). Indianapolis: Hackett.

Bandyopadhyay, P. S., & Boik, R. J. (1999). The curve fitting problem: A Bayesian rejoinder. *Philosophy of Science (Proceedings), 66*, S390–S402.

Barés Gómez, C., & Fontaine, M. (2017). *Argumentation and abduction in dialogical logic* (Magnani & Bertolotti, Eds., pp. 295–314).

Barnes, E. (1995). Inference to the loveliest explanation. *Synthese, 103*, 251–278.

Batterman, R. W. (2002). *The devil in the details: Asymptotic reasoning in explanation, reduction, and emergence*. Oxford: Oxford University Press.

Batterman, R. W., & Rice, C. (2014). Minimal model explanations. *Philosophy of Science, 81*, 349–376.

Beaulieu, Y. (2008). Peirce's contribution to American cryptography. *Transactions of the Charles S. Peirce Society, 44*, 263–287.

Beirlaen, M., & Aliseda, A. (2014). A conditional logic for abduction. *Synthese, 191*, 3733–3758.

Bird, A. (2010). Eliminative abduction – Examples from medicine. *Studies in History and Philosophy of Science, 4*, 345–352.

Blake, R. M., Ducasse, C. J., & Madden, E. H. (1960). *Theories of scientific method: The renaissance through the nineteenth century*. Seattle: University of Washington Press.

Bolzano, B. (1972). *Theory of science*. Oxford: Blackwell.

Boole, G. (1958). *The laws of thought*. New York: Dover.

Bovens, L., & Hartmann, S. (2003). *Bayesian epistemology*. Oxford: Oxford University Press.

I. Niiniluoto, *Truth-Seeking by Abduction*, Synthese Library 400,
https://doi.org/10.1007/978-3-319-99157-3

Boyd, R. (1984). The current status of scientific realism. In J. Leplin (Ed.), *Scientific realism* (pp. 41–82). Berkeley: University of California Press.

Brössel, P. (2013). Correlation and truth. In V. Karakostas & D. Dieks (Eds.), *EPSA11 Perspectives and foundational problems in philosophy of science* (pp. 41–55). Cham: Springer.

Bueno-Soler, J., Carnielli, W., Coniglio, M., & Filho, R. (2017). Formal (In)consistency, abduction, and modalities. In Magnani & Bertolotti (Eds.), pp. 315–335.

Burkhardt, R. (2013). Lamarck, evolution, and the inheritance of acquired characters. *Genetics, 194*, 793–805.

Burks, A. (1946). Peirce's theory of abduction. *Philosophy of Science, 13*, 301–306.

Burtt, E. A. (1952). *The metaphysical foundations of modern science*. New York: The Humanities Press.

Campos, D. (2011). On the distinction between Peirce's abduction and Lipton's inference to the best explanation. *Synthese, 180*, 419–442.

Carnap, R. (1962). *Logical foundations of probability* (2nd ed.). Chicago: The University of Chicago Press.

Cevolani, G. (2013). Truth approximation via abductive belief change. *Logic Journal of the IGPL*. https://doi.org/10.1093/jigpal/jzt010.

Cevolani, G., Festa, R., & Kuipers, T. (2013). Verisimilitude and belief change for nomic conjunctive theories. *Synthese, 190*, 3307–3324.

Chandler, R. (1964). The simple art of murder. In *Pearls are a nuisance*. Harmondsworth: Penguin Books.

Cleland, C. (2002). Methodological and epistemic differences between historical science and experimental science. *Philosophy of Science, 69*, 474–496.

Cleland, C. (2011). Prediction and explanation in historical natural science. *The British Journal for the Philosophy of Science, 62*, 551–582.

Climenhaga, N. (2017). How explanation guides confirmation. *Philosophy of Science, 84*, 359–368.

Cohen, L. J. (1980). What has science to do with truth? *Synthese, 45*, 489–510.

Cohen, M. R., & Nagel, E. (1934). *An introduction to logic and scientific method*. London: Routledge and Kegan Paul.

Crombie, A. (1953). *Robert Grosseteste and the origins of experimental science (1100–1700)*. Oxford: Oxford University Press.

Curd, M. (1980). The logic of discovery: An analysis of three approaches. In Nickles (Ed.) (1980a), pp. 201–219.

Currie, A. (2017). Hot-blooded gluttons: Dependency, coherence, and method in historical sciences. *The British Journal for the Philosophy of Science, 68*, 929–952.

D'Alfonso, S. (2011). Supplementing belief revision for the aim of truthlikeness. *The Reasoner, 5*, 143–145.

Day, T., & Kincaid, H. (1994). Putting inference to the best explanation in its place. *Synthese, 98*, 271–295.

de Donato Rodriguez, X. (2007). Idealization, abduction, and progressive scientific change. *Theoria: An International Journal for Theory, History and Foundations of Science, 22*, 331–338.

Descartes, R. (1954). *Geometry*. New York: Dover.

Dizadji-Bahmani, F., Frigg, R., & Hartmann, S. (2011). Confirmation and reduction: A Bayesian account. *Synthese, 179*, 321–338.

Doppelt, G. (2005). Empirical success or explanatory success: What does current scientific realism need to explain? *Philosophy of Science, 72*, 1076–1087.

Doppelt, G. (2014). Best theory realism. *European Journal of Philosophy of Science, 4*, 271–291.

Douven, I. (1999). Inference to the best explanation made coherent. *Philosophy of Science (Proceedings), 66*, S424–S435.

Douven, I. (2002). Testing inference to the best explanation. *Synthese, 130*, 355–377.

Douven, I. (2011). Abduction. In E. Zalta (Ed.), *Stanford encyclopedia of philosophy*. Stanford: Stanford University. http://plato.stanford.edu/archives/spr2011/entries/abduction/

Doyle, A. C. (1952). *The complete Sherlock Holmes*. Garden City: Doubleday.
Duhem, P. (1954). *The aim and structure of physical theory*. Princeton: Princeton University Press.
Duhem, P. (1969). *To save the phenomena: An essay on the idea of physical theory from Plato to Galileo*. Chicago: The University of Chicago Press. (Original in French in 1908).
Earman, J. (1986). *A primer on determinism*. Dordrecht: D. Reidel.
Earman, J. (1992). *Bayes or bust? A critical examination of Bayesian confirmation theory*. Cambridge, MA: The MIT Press.
Eco, U. (1990). *The limits of interpretation*. Bloomington: Indiana University Press.
Eco, U., & Sebeok, T. A. (Eds.). (1983). *The sign of three: Dupin, Holmes, Peirce*. Bloomington: Indiana University Press.
Ennis, R. (1968). Enumerative induction and best explanation. *The Journal of Philosophy, 65*, 523–529.
Fann, K. T. (1970). *Peirce's theory of abduction*. The Hague: Martinus Nijhoff.
Festa, R. (1999). Bayesian confirmation. In M. Galavotti & A. Pagnini (Eds.), *Experience, reality, and scientific explanation* (pp. 55–87). Dordrecht: Kluwer.
Fetzer, J. (1981). *Scientific knowledge: Causation, explanation, and corroboration*. Dordrecht: D. Reidel.
Fetzer, J. (1985). Science and sociobiology. In J. Fetzer (Ed.), *Sociobiology and epistemology* (pp. 217–246). Dordrecht: D. Reidel.
Fine, A. (1986). Unnatural attitudes: Realist and instrumentalist attachments to science. *Mind, 95*, 149–179.
Fitelson, B. (1999). The plurality of Bayesian measures of confirmation and the problem of measure sensitivity. *Philosophy of Science, 66*, S362–S378.
Fitzhugh, K. (2006). The abduction of phylogenetic hypotheses. *Zootaxa, 1145*, 1–110.
Fitzhugh, K. (2007). Abductive inference: Implications for "Linnean" and "Phylogenetic" approaches for representing biological systematization. *Evolutionary Biology, 35*, 52–82.
Flach, P. A., & Kakas, A. C. (Eds.). (2000). *Abduction and induction: Essays on their relation and integration*. Dordrecht: Kluwer.
Forster, M., & Sober, E. (1994). How to tell when simpler, more unified, or less *ad hoc* theories will provide more accurate predictions. *British Journal for the Philosophy of Science, 45*, 1–36.
Foster, M. H., & Martin, M. L. (Eds.). (1966). *Probability, confirmation, and simplicity: Readings in the philosophy of inductive logic*. New York: The Odyssey Press.
Frankfurt, H. G. (1958). Peirce's notion of abduction. *The Journal of Philosophy, 55*, 593–597.
Friedman, M. (1974). Explanation and scientific understanding. *The Journal of Philosophy, 71*, 1–19.
Frigg, R., Bradley, S., Machete, R., & Smith, L. (2013). Probabilistic forecasting: Why model imperfection is a poison pill. In H. Andersen et al. (Eds.), *New challenges to philosophy of science* (pp. 479–491). Dordrecht: Springer.
Fumerton, R. A. (1980). Induction and reasoning to the best explanation. *Philosophy of Science, 47*, 589–600.
Gabbay, D. M., & Woods, J. (2005). *The reach of abduction: Insight and trial*. Amsterdam: Elsevier.
Gabbay, D. M., Hartmann, S., & Woods, J. (Eds.). (2011). *Inductive logic: Handbook of the history of logic* (Vol. 10). Amsterdam: North-Holland.
Gärdenfors, P. (1988). *Knowledge in flux: Modelling the dynamics of epistemic states*. Cambridge, MA: MIT Press.
Gaultier, B. (2016). On Peirce's claim that belief should be banished from science. *Transactions of the Charles S. Peirce Society, 52*, 390–415.
Gillies, D. (1989). Non-Bayesian confirmation theory and the principle of explanatory surplus. In A. Fine & J. Leplin (Eds.), *PSA 1988* (Vol. 2, pp. 373–380). East Lansing: Philosophy of Science Association.
Good, I. J. (1960). Weight of evidence, corroboration, explanatory power, information and the utility of experiments. *Journal of the Royal Statistical Society B, 22*, 319–322.

Hacking, I. (1983). *Representing and intervening*. Cambridge: Cambridge University Press.
Hadamard, J. (1945). *The psychology of invention in the mathematical field*. Princeton: Princeton University Press. (Dover, New York, 1954).
Hajek, P. (1998). Agnosticism meets Bayesianism. *Analysis, 58*, 199–206.
Hamami, Y. (2015). The interrogative model of inquiry meets dynamic epistemic logic. *Synthese, 192*, 1609–1642.
Hanson, N. R. (1958). *Patterns of discovery*. Cambridge: Cambridge University Press.
Hanson, N. R. (1961). Is there a logic of discovery? In H. Feigl & G. Maxwell (Eds.), *Current Issues in the Philosophy of Science* (pp. 20–35). New York: Holt, Rinehart, and Winston.
Hansson, S. O. (1999). *A textbook of belief dynamics*. Dordrecht: Kluwer.
Harman, G. (1965). Inference to the best explanation. *The Philosophical Review, 74*, 88–95.
Harman, G. (1968). Enumerative induction as inference to the best explanation. *Journal of Philosophy, 65*, 529–533.
Harper, W. L. (2011). *Isaac Newton's scientific method*. Oxford: Oxford University Press.
Harrowitz, N. (1983). The body of the detective model: Charles S. Peirce and Edgar Allan Poe. In Eco & Sebeok (Eds.), *The sign of three: Dupin, Holmes, Peirce* (pp. 179–197). Bloomington: Indiana University Press.
Heath, T. L. (1956). *The thirteen books of Euclid's elements* (Vol. 1–3). New York: Dover.
Hedley, R. H. (Ed.). (1979). *Dinosaurs and their living relatives*. London: The British Museum.
Hempel, C. G. (1965). *Aspects of scientific explanation*. New York: The Free Press.
Hempel, C. G. (1966). *Philosophy of natural science*. Englewood Cliffs: Prentice-Hall.
Henderson, L. (2014). Bayesianism and inference to the best explanation. *The British Journal for the Philosophy of Science, 65*, 687–715.
Hesse, M. (1974). *The structure of scientific inference*. London: Macmillan.
Hilpinen, R. (1968). *Rules of acceptance and inductive logic* (Acta Philosophica Fennica 22). Amsterdam: North-Holland.
Hilpinen, R. (2007). On practical abduction. *Theoria: The Swedish Journal of Philosophy, 73*, 207–220.
Hintikka, J. (1968). The varieties of information and scientific explanation. In B. van Rootselaar & J. F. Staal (Eds.), *Logic, methodology, and philosophy of science III* (pp. 151–171). Amsterdam: North-Holland.
Hintikka, J. (1973). *Logic, language-games, and information: Kantian themes in the philosophy of logic*. Oxford: Oxford University Press.
Hintikka, J. (1975). Concept as vision: On the problem of representation in modern art and in modern philosophy. In *The intentions of intentionality and other new models for modalities* (pp. 223–251). Dordrecht: D. Reidel.
Hintikka, J. (1980). C. S. Peirce's 'first real discovery' and its contemporary relevance. *The Monist, 63*, 304–315.
Hintikka, J. (1997). The place of C. S. Peirce in the history of logical theory. In *Lingua Universalis vs. Calculus ratiocinator* (Selected papers 2, pp. 140–161). Dordrecht: Kluwer.
Hintikka, J. (1998). What is abduction? The fundamental problem of contemporary epistemology. *Transactions of the Charles S. Peirce Society, 34*, 503–533 (Reprinted in Hintikka, 2007, Ch. 2, with additions).
Hintikka, J. (1999). *Inquiry as inquiry: Toward a logic of scientific discovery. Selected papers V*. Dordrecht: Kluwer.
Hintikka, J. (2007). *Socratic questioning: Explorations of knowledge-seeking by questioning*. Oxford: Oxford University Press.
Hintikka, J., & Bachman, J. (1991). *What if...? Toward excellence in reasoning*. Mayfield: Mountain View.
Hintikka, J., & Hintikka, M. B. (1983). Sherlock Holmes confronts modern logic. In Eco & Sebeok (Eds.), *The sign of three* (pp. 154–169). Bloomington: Indiana University Press.
Hintikka, J., & Remes, U. (1974). *The method of analysis: Its geometrical origin and its general significance*. Dordrecht: D. Reidel.

Hintikka, J., & Suppes, P. (Eds.). (1966). *Aspects of inductive logic*. Amsterdam: North-Holland.
Hobbs, J. R. (2004). Abduction in natural language understanding. In L. Horn & G. Ward (Eds.), *The handbook of pragmatics* (pp. 724–741). Oxford: Blackwell.
Hoffmann, M. H. G. (2010). "Theoric transformations" and a new classification of abductive inferences. *Transactions of the Charles S. Peirce Society, 46*, 570–590.
Hon, G., & Rakover, S. S. (Eds.). (2001). *Explanation: Theoretical approaches and applications*. Dordrecht: Kluwer.
Hookway, C. (1985). *Peirce*. London: Routledge and Kegan Paul.
Howson, C., & Urbach, P. (1989). *Scientific reasoning: The Bayesian approach*. La Salle, IL: Open Court.
Hull, D. L. (1988). *Science as process: An evolutionary account of the social and conceptual development of science*. Chicago: The University of Chicago Press.
Husserl, E. (1913). *Ideen zu einer reinen Phänomenologie und phänomenologischen Philosophie*. English translation (1931). *Ideas: General Introduction to Pure Phenomenology*. New York: Macmillan.
Iranzo, V. (2007). Abduction and inference to the best explanation. *Theoria: An International Journal for Theory, History and Foundations of Science, 60*, 339–346.
Josephson, J., & Josephson, S. (Eds.). (1994). *Abductive inference*. Cambridge: Cambridge University Press.
Kaila, E. (1960). *Die perzeptuellen und konzeptuellen Komponenten der Alltagserfahrung*. Helsinki: Acta Philosophica Fennica 13. English translation in Kaila (1979).
Kaila, E. (1979). *Reality and experience*. Dordrecht: D. Reidel.
Kaila, E. (2014). *Human knowledge: A classic statement of logical empiricism*. La Salle: Open Court.
Kakas, A. C., Kowalski, R. A., & Toni, F. (1992). Abductive logic programming. *Journal of Logic and Computation, 2*, 719–770.
Kapitan, T. (1992). Peirce and the autonomy of abductive reasoning. *Erkenntnis, 37*, 1–26.
Kapitan, T. (1997). Peirce and the structure of abductive inference. In N. Houser, D. D. Roberts, & J. van Evra (Eds.), *Studies in the logic of Charles Peirce* (pp. 477–496). Bloomington: Indiana University Press.
Kapitan, T. (2014). Abduction as practical inference. In M. Bergman & J. Quieros (Eds.), *The commens encyclopedia: The digital encyclopedia of Peirce studies: New edition*. http://www.commens.org/encyclopedia
Ketner, K. (1985). How Hintikka misunderstood Peirce's account of theorematic reasoning. *Transactions of the Charles S. Peirce Society, 21*, 407–418.
Kitcher, P. (1981). Explanatory unification. *Philosophy of Science, 48*, 507–531.
Kitcher, P. (1993). *The advancement of science*. Oxford: Oxford University Press.
Kneale, W. (1949). *Probability and induction*. Oxford: Oxford University Press.
Konolige, K. (1996). Abductive theories in artificial intelligence. In G. Brewka (Ed.), *Principles of knowledge presentation*. Amsterdam: CSLI Publications.
Krohn, K. (1926). *Die folkloristische Arbeitsmethode: begründete von Julius Krohn und weitergeführt von nordischen Forschern*. Oslo: Aschehoug.
Kruse, F. E. (1986). Indexicality and the abductive link. *Transactions of the Charles S. Peirce Society, 22*, 435–447.
Kuhn, T. S. (1977). *The essential tension*. Chicago: The University of Chicago Press.
Kuipers, T. (1999). Abduction aiming at empirical progress or even truth approximation leading to a challenge for computational modelling. *Foundations of Science, 4*, 307–323.
Kuipers, T. (2000). *From instrumentalism to constructive realism: On some relations between confirmation, empirical progress, and truth approximation*. Dordrecht: Kluwer.
Kuipers, T. (2004). Inference to the best theory, rather than inference to the best explanation – Kinds of abduction and induction. In F. Stadler (Ed.), *Induction and deduction in the sciences* (pp. 25–51). Dordrecht: Kluwer.

Kuipers, T. (2009). Comparative realism as the best response to antirealism. In C. Glymour et al. (Eds.), *Logic, methodology, and philosophy of science 13* (pp. 221–250). London: King's College.

Kuipers, T. (2014). Empirical progress and nomic truth approximation revisited. *Studies in History and Philosophy of Science, 46*, 64–72.

Kukla, A. (1996). Antirealistic explanations of the success of science. *Philosophy of Science* 63 (*PSA 1996*, Part I, ed. by L. Darden). Chicago: Philosophy of Science Association, pp. S298–S305.

Kvanvig, J. L. (1994). A critique of van Fraassen's voluntaristic epistemology. *Synthese, 98*, 325–348.

Ladyman, J., Douven, I., Horsten, L., & van Fraassen, B. C. (1997). A defence of van Fraassen's critique of abductive reasoning: Reply to Psillos. *The Philosophical Quarterly, 47*, 305–321.

Lakatos, I. (1976). *Proofs and refutations: The logic of mathematical discovery*. Cambridge: Cambridge University Press.

Lange, M. (2004). Bayesianism and unification: A reply to Wayne Myrvold. *Philosophy of Science, 71*, 205–215.

Langley, P., Simon, H. A., Bradshaw, G. L., & Zytkow, J. M. (1987). *Scientific discovery: Computational explorations of the creative processes*. Cambridge, MA: The MIT Press.

Laudan, L. (1977). *Progress and its problems*. London: Routledge and Kegan Paul.

Laudan, L. (1980). Why was the logic of discovery abandoned? In Nickles (Ed.) (1980a), pp. 173–183.

Laudan, L. (1984). *Science and values: The aims of science and their role in scientific debate*. Berkeley: University of California Press.

Laudan, L. (1990). *Science and relativism*. Chicago: The University of Chicago Press.

Leplin, J. (1987). Surrealism. *Mind, 96*, 519–524.

Leplin, J. (2004). A theory's predictive success can warrant belief in the unobservable entities it postulates. In C. Hitchcock (Ed.), *Contemporary debates in philosophy of science* (pp. 117–132). Oxford: Blackwell.

Levi, I. (1967). *Gambling with truth*. New York: Alfred A. Knopf.

Levi, I. (1979). Abduction and demands of information. In I. Niiniluoto & R. Tuomela (Eds.), *The logic and epistemology of scientific change* (Acta Philosophica Fennica 30, pp. 405–429). Amsterdam: North-Holland.

Lipton, P. (1991). *Inference to the best explanation*. London: Routledge (Second edition in 2004).

Lipton, P. (2001a). What good is an explanation? In Hon & Rakover (Eds.), pp. 43–59.

Lipton, P. (2001b). Is explanation a guide to inference? A reply to Wesley C. Salmon. In Hon & Rakover (Eds.), pp. 93–120.

Lipton, P. (2004). *Inference to the best explanation*. London: Routledge (First edition in 1991).

Lipton, P. (2007). Alien abduction: Inference to the best explanation and the management of testimony. *Episteme, 4*, 238–251.

Liu, C. (1999). Approximation, idealization, and laws of nature. *Synthese, 118*, 229–256.

Lyons, T. D. (2003). Explaining the success of a scientific theory. *Philosophy of Science, 70*(5), 891–901.

Lyons, T. (2006). Scientific realism and the stratagema de divide et Impera. *British Journal for the Philosophy of Science, 57*, 537–560.

Mäenpää, P. (1997). From backward reduction to configurational analysis. In M. Otte & M. Panza (Eds.), *Analysis and synthesis in mathematics* (pp. 201–226). Dordrecht: Kluwer.

Magnani, L. (2001). *Abduction, reason, and science: Processes of discovery and explanation*. New York: Kluwer and Plenum.

Magnani, L. (2009). *Abductive cognition: The epistemological and eco-cognitive dimensions of hypothetical reasoning*. Heidelberg: Springer.

Magnani, L., & Bertolotti, T. (Eds.). (2017). *Springer handbook of model-based science*. Dordrecht: Springer.

Magnani, L., Li, P., & Park, W. (Eds.). (2015). *Philosophy and cognitive science II: Western and eastern studies*. Cham: Springer International Publishers.

Mayer, M. C., & Pirri, F. (1993). First-order abduction via Tableaux and Sequent Calculi. *Bulletin of the IGPL, 1*, 99–117.

McGrew, T. (2003). Confirmation, heuristics, and explanatory reasoning. *The British Journal for the Philosophy of Science, 54*, 553–567.

McKaughan, D. (2008). From ugly duckling to swan: C. S. Peirce, abduction, and the pursuit of scientific theories. *Transactions of the Charles S. Peirce Society, 44*, 446–468.

McMullin, E. (1985). Galilean idealization. *Studies in History and Philosophy of Science Part A, 16* (3), 247–273.

McMullin, E. (1992). *The inference that makes science*. Milwaykee: Marquette University Press.

Meheus, J. (2004). Adaptive logics and the integration of induction and deduction. In F. Stadler (Ed.), *Induction and deduction in the sciences* (pp. 93–120). Dordrecht: Kluwer.

Meheus, J. (2005). Empirical progress and ampliative adaptive logics. In R. Festa, A. Aliseda, & J. Peijnenburg (Eds.), *Confirmation, empirical progress, and truth-approximation* (pp. 193–217). Amsterdam: Rodopi.

Meheus, J., & Batens, D. (2006). A formal logic for abductive reasoning. *Logic Journal of the IGPL, 14*, 221–236.

Merritt, D. (2017). Cosmology and convention. *Studies in History and Philosophy of Science Part B: Studies in History and Philosophy of Modern Physics, 57*, 41–52.

Mill, J. S. (1906). *A system of logic: Ratiocinative and inductive*. London: John W. Parker. (Original edition 1843.)

Milne, P. (1996). Log[P(h/eb)/P(h/b)] is the one true measure of confirmation. *Philosophy of Science, 63*, 21–26.

Minnameier, G. (2004). Pierce-suit of truth: Why inference to the best explanation and abduction ought not to be confused. *Erkenntnis, 60*, 75–105.

Minnameier, G. (2017). Forms of abduction and an inferential taxonomy. In Magnani & Bertolotti (Eds.), pp. 175–195.

Mizrahi, M. (2012). Why the ultimate argument for scientific realism ultimately fails? *Studies in History and Philosophy of Science, 43*, 132–138.

Montague, R. (1974). Deterministic theories. In R. H. Thomason (Ed.), *Formal philosophy: Selected papers of Richard Montague*. New Haven: Yale University Press.

Monton, B. (1998). Bayesian agnosticism and constructive empiricism. *Analysis, 58*, 207–212.

Morrison, M. (2000). *Unifying scientific theories: Physical concepts and mathematical structures*. Cambridge: Cambridge University Press.

Musgrave, A. (1988). The ultimate argument for scientific realism. In R. Nola (Ed.), *Relativism and realism in science* (pp. 229–252). Dordrecht: Kluwer.

Myrvold, W. C. (1996). Bayesianism and diverse evidence: A reply to Andrew Wayne. *Philosophy of Science, 63*, 661–665.

Myrvold, W. C. (2003). A Bayesian account of the virtue of unification. *Philosophy of Science, 70*, 399–423.

Myrvold, W. (2017). On evidential import of unification. *Philosophy of Science, 84*, 92–114.

Nagel, E. (1961). *The structure of science*. New York/London: Hartcourt, Brace & World/ Routledge & Kegan Paul.

Natterer, F. (2001). *The mathematics of computerized tomography*. Philadelphia: SIAM.

Negri, S., & von Plato, J. (2001). *Structural proof theory*. Cambridge: Cambridge University Press.

Nelson, D. E. (1996). Confirmation, explanation, and logical strength. *The British Journal for the Philosophy of Science, 47*, 399–413.

Nepomuceno, Á. (2014). Scientific models of abduction: The role of non classical logic. In W. J. Gonzalez (Ed.), *Bas van Fraassen's approach to representation and models in science* (pp. 121–141). Dordrecht: Springer.

Nepomuceno-Fernández, Á., & Soler-Toscano, F. (2007). Metamodeling abduction. *Theoria: An International Journal for Theory, History and Foundations of Science, 60*, 285–293.

Nepomuceno-Fernández, Á., Soler-Toscano, F., & Velásquez-Quesada, F. R. (2017). Abductive reasoning in dynamic epistemic logic. In Magnani & Bertolotti (Eds.), *Springer handbook of model-based science* (pp. 269–293). Cham: Springer.

Nesher, D. (2001). Peircean epistemology of learning and the function of abduction as the logic of discovery. *Transactions of the Charles S. Peirce Society, 37*, 23–57.
Nickles, T. (Ed.). (1980a). *Scientific discovery, logic, and rationality*. Dordrecht: D. Reidel.
Nickles, T. (Ed.). (1980b). *Scientific discovery: Case studies*. Dordrecht: D. Reidel.
Nickles, T. (1981). What is a problem that we may solve it? *Synthese, 47*, 7–32.
Niiniluoto, I. (1972). Inductive systematization: definition and a critical survey. *Synthese, 25*, 25–81.
Niiniluoto, I. (1978). Notes on Popper as follower of Whewell and Peirce. *Ajatus, 37*, 272–327 (Reprinted in Niiniluoto (1984), Ch. 3).
Niiniluoto, I. (1981). Statistical explanation reconsidered. *Synthese, 48*, 437–472.
Niiniluoto, I. (1983). Novel facts and Bayesianism. *The British Journal for the Philosophy of Science, 34*, 375–379.
Niiniluoto, I. (1984). *Is science progressive?* Dordrecht: D. Reidel.
Niiniluoto, I. (1985). Tieteenfilosofisia näkökulmia kulttuurintutkimukseen. In J. Pennanen et al. (Eds.), *Teoriaa, aihepiirejä ja näkökulmia kansatieteellisessä tutkimuksessa* (pp. 1–18). Ethnos: Lappeenranta.
Niiniluoto, I. (1987). *Truthlikeness*. Dordrecht: D. Reidel.
Niiniluoto, I. (1988). Analogy and similarity in scientific reasoning. In D. Helman (Ed.), *Analogical reasoning* (pp. 271–298). Dordrecht: Kluwer.
Niiniluoto, I. (1990a). Measuring the success of science. In A. Fine, M. Forbes, & L. Wessels (Eds.), *PSA 1990* (Vol. 1, pp. 435–445). East Lansing: Philosophy of Science Association.
Niiniluoto, I. (1990b). Theories, approximations, idealizations. In J. Brzezinski et al. (Eds.), *Idealization I: General problems* (pp. 9–57). Amsterdam: Rodopi.
Niiniluoto, I. (1993a). Peirce's theory of statistical explanation. In E. C. Moore (Ed.), *Charles S. Peirce and the philosophy of science* (pp. 186–207). Tuscaloosa: The University of Alabama Press.
Niiniluoto, I. (1993b). The aim and structure of applied research. *Erkenntnis, 38*, 1–21.
Niiniluoto, I. (1994). Descriptive and inductive simplicity. In W. Salmon & G. Wolters (Eds.), *Logic, language, and the structure of scientific theories* (pp. 147–170). Pittsburgh: University of Pittsburgh Press.
Niiniluoto, I. (1995). Hintikka and Whewell on Aristotelian induction. *Grazer Philosophische Studien, 49*, 49–61.
Niiniluoto, I. (1997). Inductive logic, atomism, and observational error. In M. Sintonen (Ed.), *Knowledge and inquiry* (pp. 117–131). Amsterdam: Rodopi.
Niiniluoto, I. (1998). Verisimilitude: The third period. *The British Journal for the Philosophy of Science, 49*, 1–29.
Niiniluoto, I. (1999a). *Critical scientific realism*. Oxford: Oxford University Press.
Niiniluoto, I. (1999b). Defending abduction. *Philosophy of Science (Proceedings), 66*, S436–S451.
Niiniluoto, I. (1999c). Abduction and geometrical analysis. In L. Magnani, N. Nersessian, & P. Thagard (Eds.), *Model-based reasoning in scientific discovery* (pp. 239–254). New York: Kluwer and Plenum.
Niiniluoto, I. (2000). Hempel's theory of statistical explanation. In J. H. Fetzer (Ed.), *Science, explanation, and rationality: The philosophy of Carl G. Hempel* (pp. 138–163). Oxford: Oxford University Press.
Niiniluoto, I. (2004). Truth-seeking by abduction. In F. Stadler (Ed.), *Induction and deduction in the sciences* (pp. 57–82). Dordrecht: Kluwer.
Niiniluoto, I. (2005a). Abduction and truthlikeness. In R. Festa, A. Aliseda, & J. Peijnenburg (Eds.), *Confirmation, empirical progress, and truth-approximation* (pp. 255–275). Amsterdam: Rodopi.
Niiniluoto, I. (2005b). Inductive logic, verisimilitude, and machine learning. In P. Hajek, L. Valdés-Villanueva, & D. Westerståhl (Eds.), *Logic, methodology and philosophy of science: Proceedings of the twelfth international congress* (pp. 295–314). London: College Publications.

Niiniluoto, I. (2007). Structural rules for abduction. *Theoria: An International Journal for Theory, History and Foundations of Science, 22*, 325–329.
Niiniluoto, I. (2008). Abduction, unification, and Bayesian confirmation: Comment on Schurz. In C. Dégremont, L. Keiff, & H. Rückert (Eds.), *Dialogues, logics and other strange things* (pp. 365–370). London: College Publications.
Niiniluoto, I. (2009). Futures studies: Science or art? *Futura, 28*, 59–64.
Niiniluoto, I. (2011a). Abduction, tomography, and other inverse problems. *Studies in History and Philosophy of Science, 42*, 135–139.
Niiniluoto, I. (2011b). Revising beliefs towards the truth. *Erkenntnis, 75*, 165–181.
Niiniluoto, I. (2011c). The Development of the Hintikka Program. In D. Gabbay, S. Hartmann, & J. Woods (Eds.), *Handbook of the history of logic* (vol. 10: Inductive logic, pp. 311–356). Amsterdam: North-Holland.
Niiniluoto, I. (2014a). Scientific realism: Independence, causation, and abduction. In K. Westphal (Ed.), *Realism, science, and pragmatism* (pp. 159–172). New York: Routledge.
Niiniluoto, I. (2014b). Approximation of laws. In G. Guo & C. Liu (Eds.), *Scientific explanation and methodology of science* (pp. 148–155). Singapore: World Scientific.
Niiniluoto, I. (2016). Unification and confirmation. *Theoria: An International Journal for Theory, History and Foundations of Science, 31*, 107–124.
Niiniluoto, I. (2017). Optimistic realism about scientific progress. *Synthese, 194*, 3291–3309.
Niiniluoto, I. (2018). Explanation by idealized theories. *Kairos. Journal of Philosophy & Science, 20*, 43–63.
Niiniluoto, I., & Tuomela, R. (1973). *Theoretical concepts and hypothetico-inductive inference*. Dordrecht: D. Reidel.
Nowak, L. (1980). *The structure of idealization*. Dordrecht: D. Reidel.
Nyrup, R. (2015). How explanatory reasoning justifies pursuit: A Peircean view of IBE. *Philosophy of Science, 82*, 749–760.
Oddie, G. (1986). *Likeness to truth*. Dordrecht: D. Reidel.
Okasha, S. (2000). Van Fraassen's critique of inference to the best explanation. *Studies in History and Philosophy of Science, 31*, 691–710.
Olsson, E. (2005). *Against coherence: Truth, probability, and justification*. Oxford: Oxford University Press.
Paavola, S. (2004). Abduction as a logic and methodology of discovery: The importance of strategies. *Foundations of Science, 9*, 267–283.
Paavola, S. (2005). Peircean abduction: Instinct or inference? *Semiotica, 153*, 131–154.
Paavola, S. (2006). Hansonian and harmanian abduction as models of discovery. *International Studies in the Philosophy of Science, 20*, 91–106.
Paavola, S. (2015). Deweyan approaches to abduction? In U. Zachariasson (Ed.), *Activity, belief and inquiry: Pragmatist perspectives on science, society and religion* (pp. 230–249). Nordic Pragmatist Network: Helsinki.
Pappus. (1986). *Synagoge VII* (English transl. by A. Jones). New York: Springer-Verlag.
Park, W. (2017a). Magnani's manipulative abduction. In Magnani & Bertolotti (Eds.), pp. 197–213.
Park, W. (2017b). *Abduction in context: The conjectural dynamics of scientific reasoning*. Cham: Springer International Publishing.
Peirce, C. S. (1901). Modality, necessary, possibility, probability. In J. M. Baldwin (Ed.), *Dictionary of philosophy and psychology*. P. Smith: Gloucester, MA.
Peirce, C. S. (1931–35, 1958). *Collected Papers* 1–6 (C. Hartshorne & P. Weiss, Eds.), 7–8 (A. Burks, Ed.). Cambridge, MA: Harvard University Press. (*CP*).
Peirce, C. S. (1976). In C. Eisele (Ed.), *The new elements of mathematics* (Vol. I–IV). The Hague: Mouton.
Peirce, C. S. (1982–2010). In M. Fisch et al. (Eds.), *Writings of Charles S. Peirce: A chronological edition* (Vol. 1–6, 8). Bloomington: Indiana University Press. (*W*).
Peirce, C. S. (1992a). In K. L. Ketner (Ed.), *Reasoning and the logic of things: The Cambridge conferences lectures of 1898*. Cambridge, MA: Harvard University Press (*RLT*).

Peirce, C. S. (1992b). In N. Houser & C. Kloesel (Eds.), *The essential Peirce vol. 1 (1867–1893)*. Bloomington: Indiana University Press (*EP* 1).

Peirce, C. S. (1998). In N. Houser & C. Kloesel (Eds.), *The essential Peirce vol. 2 (1893–1913)*. Bloomington: Indiana University Press (*EP* 2).

Peng, Y., & Reggia, J. A. (Eds.). (1990). *Abductive inference: Models for diagnostic problem-solving*. New York: Springer.

Pietarinen, A.-V. (2006). *The signs of logic: Peircean themes on the philosophy of language, games, and communication*. Dordrecht: Springer.

Platnick, N. I., & Cameron, H. D. (1977). Cladistic methods in textual, linguistic, and phylogenetic analysis. *Systematic Zoology, 26*, 380–385.

Plutynski, A. (2011). Four problems of abduction: A brief history. *HOPOS: The Journal of the International Society for the History of Philosophy of Science, 1*, 227–248.

Poe, E. A. (1927). *Poems and essay*. London: Everyman's Library, Dent. (Reprinted, 1977).

Poe, E. A. (1976). In H. Beaver (Ed.), *The science fiction of Edgar Allan Poe*. Harmondsworth: Penguin Books.

Poe, E. A. (1977). In P. van Doren Stern (Ed.), *The portable Poe*. Harmondsworth: Penguin Books.

Poldrack, R. (2006). Can cognitive processes be inferred from neuroimaging data? *Trends in Cognitive Sciences, 10*, 59–63.

Polya, G. (1957). *How to solve it* (2nd ed.). Princeton: Princeton University Press.

Popper, K. R. (1957). *The poverty of historicism*. London: Routledge.

Popper, K. R. (1959). *The logic of scientific discovery*. London: Hutchinson.

Popper, K. R. (1963). *Conjectures and refutations*. London: Hutchinson.

Popper, K. R. (1972). *Objective knowledge: An evolutionary approach*. Oxford: Oxford University Press. (Revised edition 1979).

Psillos, S. (1996). On van Fraassen's critique of abductive reasoning. *The Philosophical Quarterly, 46*, 31–47.

Psillos, S. (1999). *Scientific realism: How science tracks truth*. London: Routledge.

Psillos, S. (2000). *Abduction: Between conceptual richness and computational complexity* (Flach & Kakas, Eds., pp. 59–74).

Psillos, S. (2001). Predictive similarity and the success of science: A reply to Stanford. *Philosophy of Science, 68*, 346–355.

Psillos, S. (2002). Simply the best: A case for abduction. In A. Kakas & F. Sadri (Eds.), *Computational logic* (pp. 605–625). Berlin: Springer.

Psillos, S. (2004). Inference to the best explanation and Bayesianism. In F. Stadler (Ed.), *Induction and deduction in the sciences* (pp. 83–91). Dordrecht: Kluwer.

Psillos, S. (2009). *Knowing the structure of nature: Essays on realism and explanation*. Houndmills: Palgrave Macmillan.

Psillos, S. (2011). An explorer upon untrodden ground: Peirce on abduction. In D. Gabbay, S. Hartmann, and J. Woods (eds.), *Handbook of the history of logic*, (vol. 10: Inductive logic, pp. 117–151). Amsterdam: North-Holland.

Putnam, H. (1978). *Meaning and the moral sciences*. London: Routledge and Kegan Paul.

Quinn, A. (2016a). William Whewell's philosophy of architecture and the historicization of biology. *Studies in History and Philosophy of Biological and Biomedical Sciences, 59*, 11–19.

Quinn, A. (2016b). Phylogenetic inference to the best explanation and the bad lot argument. *Synthese, 193*, 3025–3039.

Quinn, A. (2017). When is a cladist not a cladist? *Biology and Philosophy, 32*, 581–598.

Radon, J. (1987). In P. M. Gruber et al. (Eds.), *Collected Works* (Vol. 2). Basel: Austrian Academy of Science, Birkhäuser.

Randall, J. H. (1961). *The school of Padua and the emergence of modern science*. Padova.

Ravila, P. (1946). *Johdatus kielihistoriaan*. Helsinki: SKS.

Reichenbach, H. (1938). *Experience and prediction*. Chicago: The University of Chicago Press.

Rescher, N. (1970). *Scientific explanation*. New York: The Free Press.

Rescher, N. (1978). *Peirce's philosophy of science*. Notre Dame: Notre Dame University Press.

Roberts, R. M. (1989). *Serendipity: Accidental discoveries in science*. New York: Wiley.
Robinson, P., & O'Hara, R. J. (1992). Report on the textual criticism challenge 1991. *Bryn Mawr Classical Review, 3*, 331–337.
Roche, W., & Sober, E. (2013). Explanatoriness is evidentially irrelevant; or, inference to the best explanation meets Bayesian confirmation theory. *Analysis, 73*, 659–668.
Romdhane, L. B., & el Ayeb, B. (2011). An evolutionary algorithm for abductive reasoning. *Journal of Experimental and Theoretical Artificial Intelligence, 23*, 529–544.
Roos, T., & Heikkilä, T. (2009). Evaluating methods for computer-assisted stemmatology using artificial benchmark data sets. *Literary and Linguistic Computing, 24*, 417–433. https://doi.org/10.1093/llc/fqp002.
Rott, H. (1989). Approximation versus idealization: The Kepler-Newton case. In J. Brzezinski et al. (Eds.), *Idealization II: Forms and applications* (pp. 101–124). Amsterdam: Rodopi.
Rowbottom, D. (2015). Scientific progress without increasing verisimilitude: In response to Niiniluoto. *Studies in History and Philosophy of Science, 51*, 100–104.
Ruse, M. (1982). *Darwinism defended: A guide to the evolutionary controversies*. Reading: Addison-Wesley.
Saatsi, J. (Ed.). (2018). *The Routledge handbook of scientific realism*. London: Routledge.
Salmon, W. (1990). Rationality and objectivity in science or Tom Kuhn meets Tom Bayes. In C. W. Savage (Ed.), *Scientific theories* (pp. 175–204). Minneapolis: University of Minnesota Press.
Salmon, W. (2001a). Explanation and confirmation: A Bayesian critique of inference to the best explanation. In Hon & Rakover (Eds.), pp. 61–91.
Salmon, W. (2001b). *Reflections of a bashful Bayesian: A reply to Peter Lipton* (Hon & Rakover, Eds., pp. 121–136).
Sarkar, S. (2007). *Doubting Darwin? Creationist designs on evolution*. Oxford: Blackwell.
Shaffer, M. (2007). Idealization, counterfactuals, and the correspondence principle. In J. Brzezinski et al. (Eds.), *The courage of doing philosophy: Essays presented to Leszek Nowak* (pp. 179–204). Amsterdam: Rodopi.
Schaffner, K. F. (1980). Discovery in the biomedical sciences: Logic or irrational intuition? In Nickles (1980b), pp. 171–206.
Schippers, M. (2015). Coherence and (likeness to) truth. In U. Mäki et al. (Eds.), *Recent developments in the philosophy of science: EPSA13 in Helsinki* (pp. 3–16). Cham: Springer.
Schupbach, J. N. (2005). On a Bayesian analysis of the virtue of unification. *Philosophy of Science, 72*, 594–607.
Schurz, G. (1999). Explanation and unification. *Synthese, 120*, 95–114.
Schurz, G. (2008a). Patterns of abduction. *Synthese, 164*, 201–234.
Schurz, G. (2008b). Common cause abduction and the formation of theoretical concepts in science. In C. Dégremont, L. Keiff, & H. Rückert (Eds.), *Dialogues, logics and other strange things* (pp. 337–364). London: College Publications.
Schurz, G. (2011a). Structural correspondence, indirect reference, and partial truth: Phlogiston theory and Newtonian mechanics. *Synthese, 180*, 103–120.
Schurz, G. (2011b). Abductive belief revision. In E. Olsson & S. Enqvist (Eds.), *Belief revision meets philosophy of science* (pp. 77–104). New York: Springer.
Schurz, G. (2015). Causality and unification: How causality unifies statistical regularities. *Theoria: An International Journal for Theory, History and Foundations of Science, 30*, 73–95.
Schurz, G. (2017). Patterns of abductive inference. In Magnani & Bertolotti (Eds.), *Springer handbook of model-based science* (pp. 151–173). Dordrecht: Springer.
Sebeok, T. A. and Umiker-Sebeok, J. (1980). *"You know my method": A juxtaposition of Charles S. Peirce and Sherlock Holmes*. Gaslight Publications, Bloomington, Indiana. (Reprinted in Eco and Sebeok, 1983, pp. 11–54).
Shanahan, T. (1986). The first moment of scientific inquiry: C. S. Peirce on the logic of abduction. *Transactions of the Charles S. Peirce Society, 22*, 449–466.

Shimony, A. (1970). Scientific inference. In R. G. Colodny (Ed.), *The nature and function of scientific theories* (pp. 79–172). Pittsburgh: University of Pittsburgh Press.
Shogenji, T. (1999). Is coherence truth conducive? *Analysis, 59*, 338–345.
Shogenji, T. (2013). Coherence of the contents and the transmission of probabilistic support. *Synthese, 190*, 2525–2545.
Short, T. L. (2007). *Peirce's theory of signs*. Cambridge: Cambridge University Press.
Simon, H. A. (1977). *Models of discovery*. Dordrecht: D. Reidel.
Sintonen, M. (2009). The two aspects of method: Questioning fellow inquirers and questioning nature. In M. Sintonen (Ed.), *The Socratic tradition: Questioning as philosophy and as method* (pp. 193–221). London: College Publications.
Sloughter, D. C. (1996). Peirce and Fisher on the place of probability in abductive inference. *Philosophia Scientiae, 1*, 133–142.
Smart, J. (1989). *Our place in the universe*. Oxford: Blackwell.
Smith, L. D. (1955). *Cryptography: The science of secret writing*. New York: Dover.
Smokler, H. (1968). Conflicting conceptions of confirmation. *The Journal of Philosophy, 65*, 300–312.
Sneath, P., & Sokal, P. (1973). *Numerical taxonomy: The principles and practice of numerical classification*. San Francisco: W. Freeman and Co.
Snyder, L. J. (1997). Discoverers' induction. *Philosophy of Science, 64*, 580–604.
Sober, E. (1984). *The nature of selection: Evolutionary theory in philosophical focus*. Cambridge, MA: The MIT Press.
Sober, E. (1988). *Reconstructing the past: Parsimony, evolution, and inference*. Cambridge, MA: The MIT Press.
Sober, E. (1993). *The philosophy of biology*. San Francisco: Boulder.
Sober, E. (2008). *Evidence and evolution: The logic behind the science*. Cambridge: Cambridge University Press.
Sober, E. (2015). *Ockham's Razor: A user's manual*. Cambridge: Cambridge University Press.
Stanford, P. K. (2000). An antirealist explanation of the success of science. *Philosophy of Science, 67*, 266–284.
Stanford, P. K. (2006). *Exceeding our grasp: Science, history, and the problem of unconceived alternatives*. Oxford: Oxford University Press.
Suppes, P. (1962). Models of data. In E. Nagel et al. (Eds.), *Logic, methodology and philosophy of science: Proceedings of the 1960 international congress* (pp. 252–261). Stanford: Stanford University Press.
Tarantola, A. (2005). *Inverse problem theory*. Philadelphia: SIAM.
Tavory, I., & Timmermans, S. (2014). *Abductive analysis: Theorizing qualitative research*. Chicago: The University of Chicago Press.
Tehrani, J., Nguyen, Q., & Roos, T. (2015). Oral fairy tale or literary fake? Investigating the origins of little red riding hood using phylogenetic network analysis. *Digital Scholarship in the Humanities, 31*, 611–636. https://doi.org/10.1093/llc/fqv016.
Thagard, P. (1978). The best explanation: Criteria for theory choice. *The Journal of Philosophy, 75*, 76–92.
Thagard, P. (1981). Peirce on hypothesis and abduction. In K. Ketner et al. (Eds.), *Proceedings of the C. S. bicentennial international congress* (pp. 271–274). Texas: Texas Tech University Press.
Thagard, P. (1988). *Computational philosophy of science*. Cambridge, MA: MIT Press.
Thagard, P. (1989). Explanatory coherence. *Behavioral and Brain Sciences, 12*(03), 435–502.
Thagard, P., & Shelley, C. (1997). Abductive reasoning: Logic, visual thinking, and coherence. In M. L. Dalla Chiara et al. (Eds.), *Logic and scientific methods* (pp. 413–427). Dordrecht: Kluwer.
Tiercelin, C. (2005). Abduction and the semiotics of perception. *Semiotica, 153*, 389–412.
Truzzi, M. (1983). Sherlock Holmes: Applied social psychologist. In Eco & Sebeok (Eds.), *The sign of three: Dupin, Holmes, Peirce* (pp. 55–80). Bloomington: Indiana University Press.
Tuomela, R. (1973). *Theoretical concepts*. Berlin: Springer-Verlag.

Tuomela, R. (1985). Truth and best explanation. *Erkenntnis, 22*, 271–299.
Turner, D. (2007). *Making prehistory: Historical science and the scientific realism debate*. Cambridge: Cambridge University Press.
Tuzet, G. (2006). Projectual abduction. *Logic Journal of IGPL, 14*, 151–160.
van Fraassen, B. (1980). *The scientific image*. Oxford: Oxford University Press.
van Fraassen, B. (1989). *Laws and symmetry*. Oxford: Oxford University Press.
van Fraassen, B. (1998). The agnostic subtly probabilified. *Analysis, 58*, 212–220.
von Plato, J. (1994). *Creating modern probability*. Cambridge: Cambridge University Press.
Walton, D. (2004). *Abductive reasoning*. Tuscaloosa: The University of Alabama Press.
Wayne, A. (2011). Expanding the scope of explanatory idealization. *Philosophy of Science, 78*(5), 830–841.
Webb, J. C. (2006). Hintikka on Aristotelean constructions, Kantian intuitions, and Peircean theorems. In R. E. Auxier & L. E. Hahn (Eds.), *The philosophy of Jaakko Hintikka* (pp. 195–301). Chicago/La Salle: Open Court.
Weisberg, J. (2009). Locating IBE in the Bayesian framework. *Synthese, 167*, 125–143.
Weisberg, M. (2013). Getting serious about similarity. *Philosophy of Science, 79*, 785–794.
Whewell, W. (1847). *The philosophy of inductive sciences, founded upon their history* (2nd ed.). London: John W. Parker and Sons.
Whewell, W. (1860). *The philosophy of discovery*. London. (New Edition: 1971, Lenox Hill, Burt Franklin, New York).
Williams, M. (1982). The importance of prediction testing in evolutionary biology. *Erkenntnis, 17*, 291–306.
Williamson, T. (2016). Abductive philosophy. *The Philosophical Forum, 47*(3–4), 263–280.
Woods, J. (2007). Ignorance and semantic tableaux: Aliseda on abduction. *Theoria: An International Journal for Theory, History and Foundations of Science, 22*, 305–318.
Woods, J. (2017). Reorienting the logic of abduction. In Magnani & Bertolotti (Eds.), pp. 137–150.
Zamora Bonilla, J. P. (1996). Truthlikeness without truth: A methodological approach. *Synthese, 93*, 343–372.

Index

I. Niiniluoto, *Truth-Seeking by Abduction*, Synthese Library 400,
https://doi.org/10.1007/978-3-319-99157-3

I

J

K

L

M

MIX
Papier aus verantwortungsvollen Quellen
Paper from responsible sources
FSC® C105338

If you have any concerns about our products,
you can contact us on
ProductSafety@springernature.com

In case Publisher is established outside the EU,
the EU authorized representative is:
Springer Nature Customer Service Center GmbH
Europaplatz 3, 69115 Heidelberg, Germany

Printed by Libri Plureos GmbH
in Hamburg, Germany